RECENT ADVANCES IN FISH ECOLOGY, LIMNOLOGY AND ECO-CONSERVATION

VOLUME IV

Editor - in - Chief
SURENDRA NATH

1996

DAYA PUBLISHING HOUSE
DELHI - 110035

ISBN 81-7035-155-3

Published by : **Daya Publishing House**
1123/74, Deva Ram Park
Tri Nagar, Delhi - 110035.
Phone : 7220593
Fax : (011) 7198902

Laser Typeset by : ShiKi Associates,
New Delhi-110058

Printed at : DP's Impressive Impressions, New Delhi — 110059

PRINTED IN INDIA

To

the everlasting memory of

my esteemed grandfather

PANDIT RAGHAVENDRA NATH KOUL

the great savant of Kashmir

who initiated me into the

quest for

TRUTH

Preface

It is the aim of this new serial publication to help biologists to keep abreast of the latest developments in the different lines of research in the fields of fish ecology, limnology and eco-conservation. The present volume of Recent Advances in Fish Ecology, Limnology and Eco-conservation is the fourth one in the series and includes thirteen original research papers on the subject. The first volume was brought out in 1988 to felicitate Professor S.M. Das D.Sc., F.Z.S. (London), F.N.A.Sc., F.A.Z., F.Z.S.I., F.N.E.S.A., Scientist Emeritus, Lucknow (India) on the completion of his fifty years of active research and teaching at the Universities of Lucknow, Kashmir and Kumaun. The second volume appeared in 1992 and the third in 1994. It is intended that each volume shall contain original research papers as well as comprehensive review articles summarizing the general position of our knowledge in the vast fields of limnology, fish ecology and eco-conservation.

The Editor and the Publishers accept no responsibility for the opinions expressed by various contributors.

Surendra Nath
Editor-in-Chief

Contributions for volume V of Recent Advances in Fish Ecology, Limnology and Eco-conservation should be sent to the Editor-in-Chief at the following address:

Post Box No : 140
G.P.O. Jammu (Tawi), J. & K., 180 001
India

Contents

Contributors

1. Dr. Lubomir Hanel
   Department of Systematic Zoology,
   Charles University, Prague,
   Czechoslovakia

2. Professor (Dr.) K.C. Pandey
   Head of the Department of Zoology,
   Lucknow University, Lucknow,
   India

3. Dr. Sudish Chandra
   Department of Zoology,
   B.S.N.V. Degree College,
   Lucknow, India

4. Dr. Kaza V. Rama Rao
   Desert Regional Station,
   Zoological Survey of India,
   Jodhpur, India

5. Dr. H.A. Khan
   Freshwater Aquaculture Research
   and Training Centre,
   C.I.F.R.I., Cuttack, India

6. Dr. Tej Kumar Shreshta
   Department of Zoology,
   Tribhuvan University,
   Kirtipur Campus,
   Kathmandu, Nepal

7. Dr. Subir Kumar Mukhopadhyay
   C.I.F.R.I.,
   Barrackpore, India

8. Dr. A. Wanganeo
   Department of Limnology,
   Bhopal University,
   Bhopal (M.P.), India

9. Professor (Dr.) S.M. Das
Scientist Emeritus,
Aloka Niwas,
Lucknow, India

10. Professor (Dr.) Surendra Nath
Department of Zoology,
Camp College for Kashmir Migrants,
Government Gandhi Memorial Science College Campus,
Jammu (J. & K.), India

11. Professor (Dr.) M.K. Raina
Head of the P.G. Department of Zoology,
Kashmir University,
Srinagar (Kashmir), India

Rec. Adv. Fish Ecol. Limn. Eco-conserv.
IV : 1-5

1

Morphometrical Note on the Sucking Loach *Gyrinocheilus aymonieri* (Gyrinocheilidae; Cypriniformes)

LUBOMIR HANEL

Abstract

The author examined 29 specimens of the sucking loach *Gyrinocheilus aymonieri* (Tirant 1883) with respect to 22 plastic and 10 meristic characters, representing after Hora (1923) one of the 3 species of the family Gyrinocheilidae.

INTRODUCTION

Berg (1955) included Gyrinocheilidae as a family of the suborder Cyprinoidei (Cypriniformes). Tirant (1883, after Fowler 1938) described the new species from two localities in Cambodia as *Psilorhynchus aymonieri*. Vaillant (1902, after Berg 1906) described new genus and species *Gyrinocheilus pustulosus* from Central Borneo. On the basis of the two specimens from the region of frontiers border between Siam and Cambodia, Berg (1906) described the new species as *Gyrinocheilus kaznakovi*. Fowler (1935, 1938) and Ramaswami (1951) mutilated by mistake the species designation to *"kaznakoi"*. Hora (1935, after Fowler 1938) identified *Gyrinocheilus kaznakovi* Berg, 1906 with the species *Psilorhynchus aymonieri* Tirant, 1883. Fowler (1938) recognized two genera in the family Gyrinocheilidae, viz. *Gyrinocheilus* and *Gyrinocheilops*, the last with a new species *Gyrinocheilops pennocki*. Data concerning the reproduction in captivity are missing and *Gyrinocheilus aymonieri* is kept in aquaria only due to new imports, even of very small specimens (about 35 mm.) The detailed study of this fish is, therefore, interesting, while all specimens kept in tanks of several Czech aquarium hobbyists examined by the present author originate from nature, but without designation of the locality.

MATERIALS AND METHODS

The author had at his disposal a total of 27 specimens (35-50 mm. in body length, 1 specimen 107 mm. and 1 specimen 121 mm. in body length) of *Gyrinocheilus aymonieri*. The material was preserved at first in formalin and later transferred to 80 per cent spirit. All measurements were made with the accuracy of ± 0.5 mm., only eye diameter, length of snout, width of snout, length of transverse groove on throat, interorbital distance and depth of caudal peduncle was measured with the accuracy of ± 0.1 mm. The number of scales was counted by the binocular microscope (magn. 17-24 X); the number of gill-rakers was counted using the monocular microscope.

RESULTS AND DISCUSSION

A comparison of the results obtained by the present author with Berg's (1906) data (given in brackets) revealed an altogether agreement of values in plastic characters: in per cent of body length - head length 23.3 - 28.6, average 25.9 (21.7); in per cent of body length - body depth 17.0 - 23.1, average 18.5 (18.7); in per cent of body length - eye diameter 4.1 - 7.1, average 5.7 (3.0); in per cent of body length - snout length 10.5 - 13.2, average 11.8 (12.6); in per cent of body length - length of caudal peduncle 9.6 - 17.8, average 15.0 (19.9); in per cent of body length - basis of dorsal fin 17.1 - 25.0, average 19.8 (20.8); in per cent of body length - basis of anal fin 5.2 - 8.9, average 6.8 (7.2); in per cent of body length - length of pectoral fin 17.8 - 25.3, average 21.2 (24.7); in per cent of body length - length of ventral fin 14.0 - 19.7, average 17.2 (17.1); in per cent of body length - length of caudal fin 20.6 - 29.0, average 26.3 (19.3); in per cent of eye diameter - length of transverse groove on throat. This groove begins at the point of junction of the suboperculum, preoperculum and interoperculum and continues transversely on the snout (139.0 - 163.6, average 153.3), (150.0). Differences in data are due to measuring of smaller specimens than examined by Berg (1906), who had at his disposal only two specimens (the bigger one measured 166 mm.). The biggest difference was found to be in eye diameter in per cent of the body length. When this value was compared in the material with the present author in 6 specimens of 35 - 40 mm. body length, it was found that the eye diameter was 5.9 - 7.1 per cent, average 6.4 per cent of the body length, and in 2 specimens of 107 - 121 mm. of body length this value was found to be 4.1 - 4.7 per cent, average 4.4 per cent. Lower value in length of caudal fin in per cent of body length is due to the smaller size of the specimens, while the relationship of the length of caudal fin to body length reduced relatively with the increase of body length. In the specimens under report, in length group of 35 - 40 mm. body length, this value was found to be 26.2 - 29.0, average 27.9, and in specimens of 107 - 121 mm. body length it was 20.6 - 20.7, average 20.7.

Further plastic characters in the material under report which had not been mentioned by Berg (1906) were found to be as follows : In per cent of head length - snout length 37.9 - 52.6, average 44.4; in per cent of head length - eye diameter 17.2 - 25.5, average 22.1; in per cent of snout length - eye diameter 33.3 - 62.5, average 48.6; in per cent of interorbital distance - eye diameter 35.7 - 71.4, average 56.4; in per cent of head length - mouth width 12.5 - 24.1, average 18.3; in per cent of head length - interorbital distance 32.6 - 48.3, average 38.3; in per cent of length of pectoral fin - length of cutaneous pad in axilar part of the pectoral fin 20.0 - 36.0, average 27.5; in per cent of length of ventral fin - length of axillary scale of the ventral fin 20.0 - 35.0, average 27.6; in per cent of head length and length of the first branched ray in dorsal fin 66.7 - 91.3, average 79.6; in per cent of head length and length of first branched ray in anal fin 60.9 - 91.3, average 66.8; in per cent of head length - depth of caudal peduncle 29.0 - 51.7, average 35.6; in per cent of head length and length of pectoral fin 75.0 - 95.7, average 82.2; in per cent of length of ventral fin - distance of vent to extremity of ventral fin 30.0 - 61.3, average 45.7; in per cent of head length - distance from end of snout to centre of eye 49.6 - 68.6, average 56.3; in per cent of head length - head width 50.0 - 75.9, average 56.8; in per cent of length of basis - dorsal fin distance from beginning of dorsal fin to vertical line from basis of ventrals 44.4 - 87.5, average 64.8.

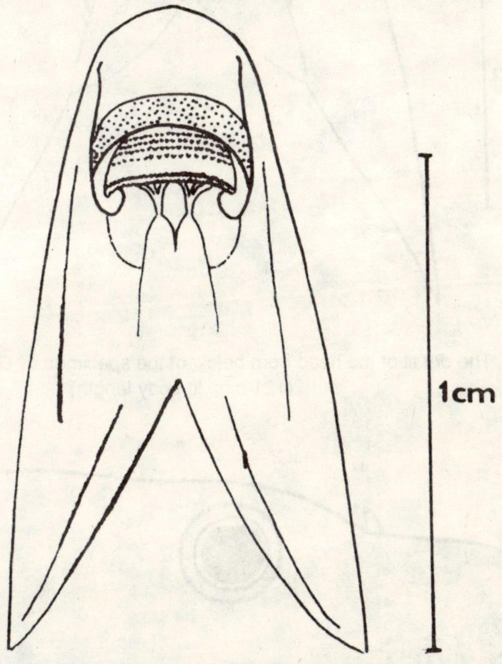

Fig. 1 : The detail of the head from below of the specimen of *Gyrinocheilus aymonieri* (45 mm. in body length).

When the present author compared the meristic characters with Berg's (1906) description, he found that the number of fin rays in dorsal and anal fins did not agree. Berg (l.c.) cited for dorsal fin the following formula: III/9, while the present value was found to be II/9/I. In the anal fin Berg found II/5, while the author's value was found to be II/4-5/I, average II/5/I. In both fins the author found the Ist ray unbranched throughout its entire length, while the preceding ray was branched. This fact may easily be overlooked. It is interesting that Fowler (1938) also found in anal and dorsal fins the last unbranched fin ray and cited for the species *Gyrinocheilops pennocki* with the anal fin formula; III/5/I and for the dorsal fin II/10/I. Berg (l.c.) found 39-40 scales in lateral line, while in the present material there were 39-43 (average 41) scales in the lateral line.

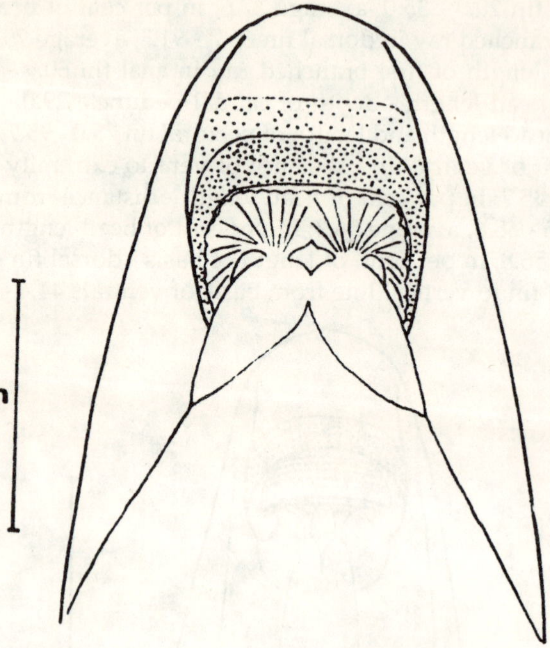

Fig. 2 : The detail of the head from below of the specimen of *Gyrinocheilus aymonieri* (121 mm. in body length).

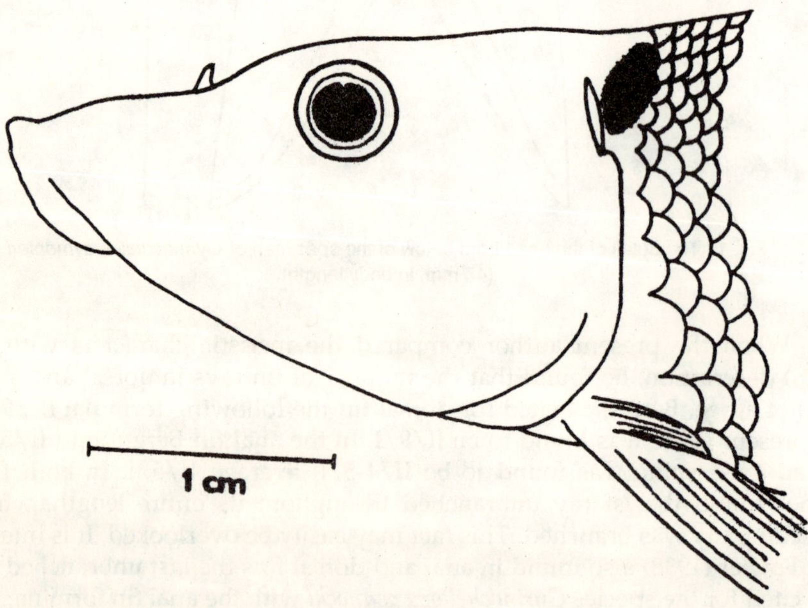

Fig. 3 : The detail of the head from the side of a specimen of *Gyrinocheilus aymonieri* (121 mm. in body length).

Above the lateral line Berg recorded 6 scales, while in the material under report there were found 6-8, average 7 scales.

Further meristic characters in the present material which Berg (l.c) did not mention are : number of rays in pectoral fins 1/11-12/1, average 1/12/1; number of rays in pelvic fins 1/7/1; scales below lateral line 5-6, average 5; scales below lateral line to anal fin 4-6, average 5; scales in predorsal region 22-27, average 25; number of gill rakers about 140.

The validity of the genus *Gyrinocheilops* (sensu Fowler 1938) is doubtful, because differences between the two genera are doubtful (see colouration). Also, the position of eye as generic character cannot be used. According to the present observations, value in per cent of head length - distance from end of snout to centre of eye is very variable. Fowler (l.c.) observed differences (*Gyrinocheilus*: eye well postmedian or near last third in the head; *Gyrinocheilops*: eye slightly postmedian in the head) which are not confirmed by the present author. As regards the colouration of the examined specimens, this agrees with Berg's (1906) description and with Fowler's (1938) figure of the species *Gyrinocheilus kaznakovi*.

SUMMARY

The author examined 29 preserved specimens of *Gyrinocheilus aymonieri* originating from aquaria of native fanciers. Plastic characters altogether agree with Berg's (1906) description; eventual differences are caused by smaller fish in the material. Concerning meristic characters the author found differences with Berg's (l.c.) description in the number of branched and unbranched rays in dorsal (present formula : II/9/I) and anal (present formula : II/4-5/I) fins. Concerning the number of scales in the lateral line and above the lateral line, the present observations agree with those of Berg (l.c.)

ACKNOWLEDGEMENTS

Thanks are due to Asstt. Professor Dr. O. Oliva who placed all the material at my disposal. Material of examined specimens was donated to the Department of Systematic Zoology by the courtesy of Drs. S. Frank, M. Chvojka, I. Petrovicky, K. Polak and R. Zukal.

REFERENCES

Berg, Leo S. 1906. *Trudy S : Peterburskogo obsc. estestvoispyt.*, 37 (7 - 8) : 1 - 4.

Berg, Leo S. 1955. Sistema ryboobraznych i ryb nyne zivuscich i iskopaemych. *Trudy zool. inst.*, 20 : 199.

Fowler, H.W. 1935. Zoological results of the third de Schauensee Siamese expedition, Part I. Fishes. *Proc. Acad. Nat. Sci., Philadelphia*, 86 : 137

Fowler, H.W. 1938. Zoological results of the third de Schauensee expedition, Part VIII : Fishes obtained in 1936. *Ibid.*, 89 : 158-160.

Ramaswami, L.S. 1951. Skeleton of cyprinoid fishes in relation to phylogenetic studies, I : The systematic position of the genus *Gyrinocheilus* Vaillant. *Proc. Nat. Inst. Sci. India*, 18(2): 125-140.

Rec. Adv. Fish Ecol. Limn. Eco-conserv.
IV : 6-9

2

On a New Cestode *Paricterotaenia dasi* n.sp. from *Burhinus oedicnemus* (Linn.) at Faizabad

K.C. PANDEY and VARSHA TAYAL

INTRODUCTION

Out of 52 specimens of the Stone curlew, *Burhinus oedicnemus* (Linn.) examined from district Faizabad, three were found infected with about 25 cestodes, belonging to the genus *Paricterotaenia* Fuhrmann, 1932, the site of infection being the intestine. Since these specimens differ from all the known species in some salient features, it is described herein as a new species. The species has been named after Prof. S.M. Das.

MATERIALS AND METHODS

Live cestodes were collected from the alimentary canals of avain host and fixed in 5 per cent formalin, stained with aceto-alum carmine, cleared in xylol and mounted in canada balsam. Sections were cut by usual microtomy technique at 8 micron thickness, stained with haematoxylin and eosin. Drawings were made with the aid of camera lucida and measurements with occulometer.

DESCRIPTION

Body measured 10-14 mm. × 0.40-0.45 mm. Scolex almost rounded and, measured 0.40-0.45 mm. × 0.54-0.59 mm. in size. Small cuticular spines present at the tip of rostellum; rostellum eversible, armed with a single row of 10 sickle-shaped hooks. Hooks measured 0.07-0.08 mm. × in length, handle 0.03-0.04 mm., blade 0.04-0.05 mm. and guard 0.01-0.02 mm. in size. Suckers round to oval, unarmed, muscular and measured 0.25-0.30 mm × 0.23-0.30 mm. in size. Neck absent. Proglottids broader than long and craspedote. Mature proglottids measured 0.17-0.20 mm × 0.30-0.39 mm. in size. Gravid proglottids 1.05-1.20 mm × 0.30-0.45 mm. in size. Longitudinal muscle layers well developed and arranged in two distinct layers. Oblique and transverse muscles scattered in parenchyma, were also observed.

Testes round to oval, 20-27, and situated in postovarian field. They measured 0.01-0.2 mm. × 0.02-0.03 mm. in size. Cirrus sac tubular, extending a little beyond the excretory canal and located in anterior half of the proglottid. Internal and external seminal vesicles present. Genital pores alternate regularly.

Ovary symmetrically lobed, present in anterior half of the proglottid and measured 0.03-0.04 mm. × 0.06-0.08 mm. in size. Vitelline gland oval, compact, located just posterior to ovary and measured 0.01-0.015 mm. × 0.020-0.025 mm. in size. Vagina narrow and opening posterior to the cirrus sac in the genital atrium. Receptaculum seminis present. Uterus sac-like and finally breaking into egg capsules, containing onchosphere. Egg capsules round to oval and measuring 0.065-0.069 mm. × 0.043-0.048 mm. in size. Onchosphere with six hooks and measuring 0.015-0.017 mm. × 0.010-0.012 mm.

Dorsal longitudinal excretory canal present. Ventral longitudinal canal broader than dorsal canal and measuring 0.010 mm. - 0.019 mm. in breadth. A transverse excretory canal joins the two ventral longitudinal canals. Genital ducts passing in between the longitudinal excretory canals.

DISCUSSION

López-Neyra (1952) has considered *P. porosa*, the type species of *Paricterotaenia* Fuhrmann, 1932 with developing uterine capsules as a member of the genus *Choanotaenia* Raillet, 1896. He has, therefore, considered *Paricterotaenia* as a synonym of *Choanotaenia*. Yamaguti (1959), however, in his classical work "Systema Helminthum" has retained the genus *Paricterotaenia* as a valid genus and has been followed by Spasski and Spasskaya (1965). The authors agree with Yamaguti (1959) and Spasski and Spasskaya (1965) and retain the genus *Paricaterotaenia* as a valid genus.

As far as the present authors are aware, following species of *Paricterotaenia* have been described from India so far, viz., *P. coronata* (Creplin, 1829); *P. decacantha* (Fuhrmann, 1913); *P. magnicirrosa* (Meggitt, 1926); *P. cirrospinosa* Patwardhan, 1935; *P. milvi* Singh, 1952; and *P. sujarensis* Mukherjee, 1970. It comes closer to *P. coronata, P. milvi, P. decacantha* and *P. cirrospinosa* in having 10 rostellar hooks arranged in a single row. The species described herein differs from all these in the size of hooks and number of testes. It differs from *P. sujarensis* in number and size of hooks and number of testes.

Among the species described from foreign lands it comes closer to the following species in having 10 rostellar hooks. *P. arquata* (Clerc, 1906); *P. rotunda* (Clerc, 1911); *P. chionis* (Fuhrmann, 1921); *P. decacantha* (Fuhr, 1913); *P. dodecacantha* (Karabbe, 1869); *P. dominicanus* (Railliet et Henry, 1912); *P. producta* (Krabbe, 1896); and *P. stellifera* (Krabbe, 1896). It differs from all the above, mentioned species in size of body and rostellar hooks, in the number of testes and presence of small cuticular spines on rosteller sac tip.

ACKNOWLEDGEMENTS

We are grateful to Prof. B.K. Tandon for laboratory facilities, to Dr. S.C. Baugh for valuable suggestions and to Sri P.N. Pande for providing the avain host. One of us (V.T.) is thankful to C.S.I.R., New Delhi for financial help.

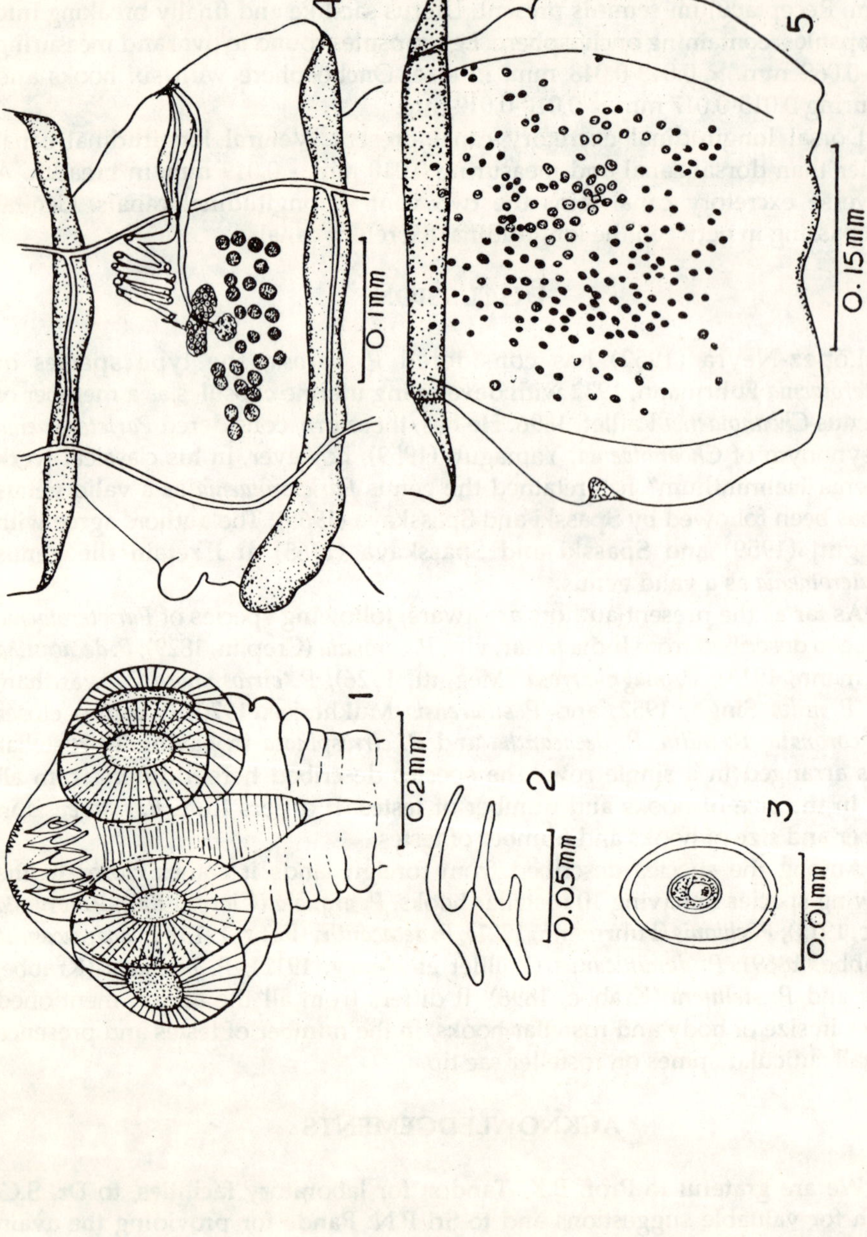

Fig. 1 : Scolex; Fig. 2 : Rostellar hooks; Fig. 3 : Egg from gravid proglottid; Fig. 4 : Mature proglottid; Fig. 5 : Gravid proglottid

REFERENCES

Clerc, W. 1906. Note sur les cestodes d'oiseaux de l'Oural. I. *Centrabl. Bakt.I.Abt.*, 42: 433-436.

————. 1911. Liste des oiseaux ouverts durant les excursions de 1905 dans la region de l'Oural antre 54° et 57° L:N. afin d'en etudier les parasites. *Bull.Soc. Ouralienne Sc. Nat.*, 30: 91-98.

Creplin, F.C.H. 1829. Novae observations de Entozois. *Berolin*, 133.

Fuhrmann, O. 1913. Nordische Vogelcestodes aus dem Muswaum Goteborg, Medd. *Goteborg Musci. Zoologiska, Afdelning*, 141.

Fuhrmann, O. 1920. Considérations générals sur les Davainea'. *Festschrift für Zschokke, Bale*, 19.

Fuhrmann, O. 1932. Les Ténias des oiseaux. *Mem. Univ. Neuchatal*, 8: 381.

Krabbe, H. 1869. Bidrag til KundsKab om Fuglenes Baendelorme. Kgl. Danske Videnska. Sclskab, Skrifter Naturvidenskab. *Math. Afdel.* 8: 149-363.

López-Neyra, C.R. 1952. Análisis Critico de los género *Choanotaenia, Anomotaeniaya* fines con redescripción de la *Taenia porosa* Rud. 1810 e invalidez del género *Paricterotaenia* Fuhrmann, 1932. *Rev. Ibér. Par.*, 12: 1-58.

Meggitt, F.J. 1926. The tapeworms of the domestic fowl. *J. Burma Research Soc.*, 15: 222-243.

Mukherjee, R.P. 1970. "Fauna of Rajasthan" India. part. 9 Cestoda. *Rec. Zool. Surv. India*, 62: 191-215.

Patwardhan, S.S. 1935. On two new species of cestodes from a snipe. *Zool. Jahrb. Syst.*, 66 : 541-548.

Railliet, A. and Henry, A. 1912. Helminths reculillis par l'expedition antartique francaise du Pourquoi-Pas. I Cestodes d'oiseaux. *Bull. Musee. Nat. Hist. Nat.* 18 : 35-39.

Singh, K.S. 1952. Cesotde parasites of birds. *Ind. J. Helminth.*, 4 : 1-72.

Spasski, A.A. and Spasskaya, L.P. 1965. Revision of the gneus *Paricterotaenia* (Cestoda: Dilepidisae) in (Parasites of animal and plants) Kishinov. *"Kartya moldovenyaski"*, 1 : 84-103.

Yamaguti, S. 1959. *Systema Helminthum*, Interscience Publishers. Vol. II : 1-769.

Rec. Adv. Fish Ecol. Limn. Eco-Conserv.
IV : 10-13

3

Sexual Variations in Serum Glutamic-Pyruvic-Transaminase Levels of Freshwater Teleost Fishes

SUDHISH CHANDRA

INTRODUCTION

It has been observed by different workers that various body and blood constituents of fishes differ in the two sexes, since biochemical constituents of the body are related to maturity, breeding cycles and varying growth rates of the fish. Sexual maturation is accompanied by profound changes in the chemistry of the fish and differences are likely to disappear at times of the year when the gonads are inactive, but several constituents change vigorously at the onset of maturity.

Transamination represents one of the principal metabolic pathways for the synthesis and deamination of aminoacids. Glutamic-Pyruvic-Transaminase is one of the common transaminases, which actively takes part in transamination reactions of the body. Transaminase levels of fishes have been studied by several workers, e.g. Creach 1967, Bell 1968, Furia *et al.* 1973 and Wilson 1973. Sex-related trends in various blood constituents have been observed by different workers in different parts of world (Alm 1959, Robertson *et al.* 1961, Booke 1964, Kirsipuu 1964, McCartney 1965, 1966, Yamashita 1967, Love 1970 and others). In this study the sexual variations in serum glutamic-pyruvic-transaminase levels (S.G.P.T.) were observed in 19 species of 14 genera and 10 families of the class Teleostomi.

MATERIALS AND METHODS

Live fishes were obtained from river Gomati and ponds from Lucknow and its suburbs. Method of catching, transport, maintenance of fishes and collection of blood etc. were the same as described earlier (Tandon and Chandra 1976, 1977). Method of King and Wootton (1959) was followed for the determination of S.G.P.T. levels. Optical density was measured at 505 nm, using Bausch and Lomb Spectronic-20 colorimeter.

RESULTS

Sex-related trends were markedly evident in the S.G.P.T. levels of all the fishes examined. The maximal difference of 67.7 per cent between the mean serum-enzyme levels of two sexes was seen in the carp *Labeo gonius*, while the minimal sexual difference of 0.20 per cent only was found in catfish *Mystus menoda*. Out of the 19 species of freshwater teleosts examined during these studies, the females of 12 species, viz., *Notopterus chitala, N. notopterus, Labeo gonius, L. calbasu, Waliago attu, Heteropneustes fossilis, Clarias batrachus, Aorichthys seenghala, Rita rita, Channa marulius, C. punctatus,* and *Anabas testudineus* had higher S.G.P.T. levels than their males. Rest of the 7 species of fishes had lower S.G.P.T. levels in females than in their males. The two species of featherbacks (*N. chitala* and *N. notopterus*) showed sexual differences of 1.14 per cent and 3.06 per cent respectively. Among carps *C. mrigala* and *C. catla* had higher serum-enzyme level in males than in their females, with differences of 25.7 per cent and 54.3 per cent in the two sexes respectively. The females of the freshwater shark *Wallago attu*, stinging catfish *Heteropneustes fossilis* and freshwater catfish *Clarias batrachus* had 12.00 per cent, 37.47 per cent and 0.80 per cent higher levels than their males. *Aorichthys seenghala* and *Rita rita* females with higher enzyme levels showed 12.30 per cent and 17.80 per cent difference respectively. The fish *Bagarius bagarius* had a difference of 42.50 per cent between the enzyme level of the two sexes. Among murrels, *Channa marulius* and *C. punctatus* revealed lower S.G.P.T. levels in males than in females with differences of 23.40 per cent and 13.50 per cent respectively, while *C. striatus* males had 35.00 per cent higher enzymatic activity. Among perches, *Nandus nandus* males revealed 9.40 per cent high enzyme levels, while the females of *Anabas testudineus* had 20.20 per cent higher enzyme levels. The sexual difference in the spiny eel *Mastacembelus armatus* was 7.30 per cent only.

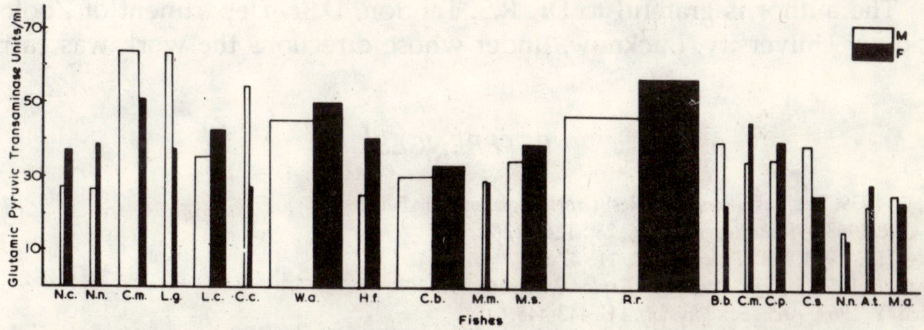

Fig. 1 : S.G.P.T. levels in the two sexes of 19 species of freshwater teleost fishes

Statistically the sexual differences in the S.G.P.T. levels in *W. attu, C. batrachus, L. clabasu, R. rita, N. chitala, N. notopterus, M. menoda, A. seenghala, C. punctatus, N. nandus* and *M. armatus* were insignificant (P>0.01), while in the rest of 8 species of fishes the differences were statistically significant (P<0.01).

DISCUSSION

Sex-related trends have been observed in the various physical, chemical and biochemical parameters of fishes by different workers (Hoar 1957,

Grant *et al.* 1970, Fletcher *et al.* 1975). The enzyme levels in the two sexes vary in fishes due to variation in sizes of the male and females (Nikolsky 1963), different maturation time of males (Chrzan 1950, Robertson *et al.* 1961, McCartney 1966), longer life of females (Hoar 1957) and greater stamina of males (Pyle 1965). In the 19 species of fishes examined here, the enzyme levels varied between different species, as well as between two species of the same genus. The general range of S.G.P.T. levels in female fishes of ten families varied from 8.50 units/ml in *M. armatus* to 141.42 units/ml in *R. rita* and interestingly the lowest and the highest enzymatic activities were observed in the female sex among the 19 species examined. Among the males of the 19 species of fishes the general minimal level (10.26 unit/ml) was found in *A. seenghala* and the maximal level (115.40 unit/ml) in *C. mrigala*. Thus the minimal general level was slightly higher in the male while the maximal level was higher in the female sex, but the fishes belonged to entirely different families and groups.

Fletcher *et al.* (1975) has mentioned that differences in protein levels exist in the two sexes of salmon of the same freshwater spawning ground while Onishi *et al.* (1969) found differences in the enzyme levels of two sexes of salmonid fishes. Interestingly, however, out of the four carps examined, three of them revealed significant differences in the S.G.P.T. levels of the two sexes. It is also of interest to note that the females of almost all active predatory fishes with accessory respiratory organs, such as *C. batrachus, H. fossilis, A. testudineus, R. rita, C. punctatus* and *C. marulius*, had higher S.G.P.T. levels than their males.

ACKNOWLEDGEMENTS

The author is grateful to Dr. R.S. Tandon, D.Sc. Department of Zoology, Lucknow University, Lucknow, under whose directions the work was carried out.

REFERENCES

Alm, G. 1959. *Rep. Inst. Freshwat. Res. Drottningholm.*, 40 : 5-145.

Bell, G.R. 1968. *J. Fish. Res. Bd. Can.*, 25 : 1247-1268.

Booke, H.E. 1964. *N.Y. Fish Game J.*, 11 : 47-57.

Chrzan, F. 1950. *J. Cans. Perm. Int. Explor. Mer.*, 16 : 192-207.

Creach, Y. 1967. *Arch. Sci. Physiol.*, 11 : 443-448.

Fletcher, G.L., Watts, E.G. and King, M.J. 1975. *J. Fish. Res. Bd. Can.*, 32 : 78-82.

Furia, M., Lamantia, G. and Scardi, V. 1973. *Boll. Soc. Ital. Soc. Ital. Biol. Sper.* 49 : 1433-1437.

Grant, B.F., Mehrle, P.M. and Russle, T.R. 1970. *Camp. Biochem. Physiol.*, 37 : 321-330.

Hoar, W.S. 1957. In : *The Physiology of Fishes*, ed. M.E. Brown, Vol. I. Academic Press, New York.

King, E.J. and Wootton, I.D.P. 1959. *Micro-analysis in Medical Biochemistry*, 3rd ed., J. and A. Churchill Ltd., London.

Kirsipuu, A. 1964. *Izu. Akad. Nauk. Eston. SSR. Biol. Ser.*, 278-283.

Love, R.M. 1970, *The Chemical Biology of Fishes*, Academic Press, New York.

McCartney, T.H. 1965. *Fish. Res. Bull. N. Y.*, 28 : 42-43.

McCartney, T.H. 1966. *Fish. Res. Bull. N.Y.*, 29 : 72-75.

Nikolsky, G.V. 1963 : *The Ecology of Fishes*, Academic Press, New York.

Onishi, T. and Murayama, S. 1959. *Bull. Tokai Reg. Fish. Res. Lab.*, 59 : 111-119.

Pyle, E.A. 1965. *Fish. Res. Bull. N.Y.*, 28 : 55-59.

Robertson, O.H., Krupp, M.A., Thomas, S.F., Home, S., Favour, C.B. and Wexler, B.C. 1961. *Gen. Camp. Endocrinal.*, 1 : 473-484.

Tandon, R.S. and Chandra, S. 1976. *Z. Tierphysiol. Tierernah. V. Futtermittelkde.*, 37 : 330-333.
Tandon, R.S. and Chandra, S. 1977. *Z. Tierphysiol. Tierernah. V. Futtermitellkde.*, 38 : 211-214.
Wilson, R.P. 1973. *Camp. Biochem. Physiol. B. Camp. Biochem. Physiol. B. Camp. Biochem.*, 46 : 617-624.
Yamashita, H. 1967. *Bull. Jap. Soc. Scient. Fish.*, 33 : 195-203.

Rec. Adv. Fish Ecol. Limn. Eco-conserv.
IV : 14-20

4

Larval Fish Taxonomy

KAZA V. RAMA RAO and S. BASKARAN

Abstract

Characters useful in the identification of fish larvae are given. Meristic characters are more important than morphometric characters. Among meristics myomerevertebrae and caudal fin elements are found to be more useful in identification. Pigmentation greatly aids in identifying early larval stages.

INTRODUCTION

Scant attention has been paid by the systematic ichthyologists to larval characters which can aid significantly in differentiating taxa and in defining evolutionary lineages within a family or order (Ahlstrom and Moser 1976). The mortality that occurs in the larval stages of commercially important fishes will affect the future adult population which will in turn affect the annual total fish landings. To understand this mortality rate the fish larvae should be identified and studied. There are two basic approaches to the study of the fish larvae : either the larval forms that are collected during ichthyoplankton surveys are studied by comparing the characters of both larva and adult (taxonomic approach) or the fertilised eggs that are taken from known parents are reared through embryonic and larval stages to the juveniles (experimental approach).

Ichthyoplankton Surveys

Eggs, larvae and juveniles of marine fishes collected by plankton gear are commonly called the ichthyoplankton. The collections are usually made on survey cruises or on wide ranging expeditions to study the fish resources.

Collection and Preservation of Ichthyoplankton

During collection the entire depth range of the eggs and larvae of the fishes to be investigated should be covered in order to obtain information on availability throughout their distributional range. The larval forms that are collected should be preserved immediately; even a time delay of few minutes will destroy the quality of the sample (Ahlstrom 1976). Usually formalin (formaldehyde-CH_2O) and ethanol (ethyl alcohol - CH_3CH_2OH) are used as preservatives. Ethanol not only causes shrinkage, desiccation and distortion but also evaporates rapidly, thus exposing the specimens to deterioration; hence it is not suitable. Richards and Berry (1973) recommended 5 to 10 per cent formalin. The authors are using 10 per cent formalin (1 part to 9 parts of water) to preserve the larvae and they are preserved in good condition. Since formalin is acidic, a buffer should be added to produce a neutral or slightly basic solution. Five grams of Borax (Sodium tetraborate) in one litre of 10 per cent formalin will provide immediate buffering, so that the sample can be stored effectively for 4 to 5 years (Richards and Berry 1973). When 1 ml. of IONOL CP-40, is added to an 8 ounce sample jar of plankton and protected from light, it will retain the red pigments in fish larvae for a period of three years. The container should be filled completely with preservative to avoid loss by evaporation and should be stored away from heat.

Zooplankton Volume Determination

Volume determination will be helpful in measuring the zooplanktonic biomass roughly. Before volume determination, the plankton is separated from the preservative by pouring the sample through a fine-meshed draining cone. The volume of the plankton thus obtained is determined by measuring the change in volume of a partly-filled graduated cylinder on its addition (Displacement volume). During volume determination there is a necessity to remove the larger forms. Richards and Berry (1973) recommended retention of larger fishes with the larvae, because they often aid in larval identification.

Sorting

Before sorting, organisms other than fish larvae should be separated. While sorting it is best to have a small sample instead of large amounts of plankton at one time. Since the fish larvae are fragile, eyedroppers, wire loops or micordissection forceps should be used. After sorting, the larvae can be stored in small vials containing 10 per cent solution of buffered formalin. Then from these samples the larvae can be taken and studied for identification.

Characters used in Identification of Larvae

Generally larval forms of closely related species and sometimes genera and even families look alike. At the same time, the larvae of distantly related forms may be closely similar in gross appearance. The larval forms that are collected during ichthyoplankton surveys will be in different developmental stages. On the basis of the development of the caudal fin, the larval forms of teleosts can be divided into three substages (Moser *et al.* 1977):

(1) Preflexion

The substage beginning with the termination of yolksac stage and ending with the initiation of notochord flexion.

(2) Flexion

The substage from the initiation of notochord flexion to its completion. At this substage the posterior edge of the hypural plate arrives at a vertical position.

(3) Postflexion

From the flexion substage to the initiation of transformation. (The change from larva to juvenile is termed as transformation or transition.)

To establish life history series through fish larvae taken in ichthyoplankton surveys, it is necessary to tie larval series to juveniles or adults, primarily through meristic characters and secondarily through morphometric characters. So the ichthyoplanktonologist has to be thoroughly acquainted with the meristic and morphometric characters of the adult fishes.

A. Meristic Characters

Meristic characters are given more importance than morphometric characters in identification because meristic characters develop early in the larval stage and they will not change in the later life stages. On the other hand the morphometric characters will change markedly between larval and later life stages. The following meristic characters are used in identification.

(1) Myomere-Vertebrae

Myomeres begin forming in the embryonic stage and the formation of the adult complement is complete during the larval stage. So the counting of myomere-vertebrae (abdominal and caudal) will be helpful to identify the larvae. Partial counts of myomeres can be used as reference points to locate the position of the origin of the fins and their migration during growth. The arrangement of dorsal and anal fin pterygiophores in relation to neural and haemal spines will be of much use in identification.

(2) Fins

The sequence of formation of various fins, the number of spines and rays and their relative length in each fin, the position of the fins on the body and migration of some fins are indispensable in identifying the larvae.

(a) Pectoral fin

The size, shape, depth of base and pigment pattern of pectoral fin are helpful in identifying the larvae up to genera and species level.

(b) Pelvic fin

The position of the pelvic fin on the body whether abdominal, thoracic or jugular is quite helpful in identifying the larval forms. The number of rays and spines of the pelvic fin will be helpful in identification up to family level.

(c) Dorsal and Anal fins

The number of spines in the dorsal and anal fins is generally constant in many families and genera and to some extent the soft-rays. Since the spines and rays of the dorsal and anal fins are forming fairly early in the larvae, they are useful in identification. In many species the terminal soft-ray of dorsal and anal fins is bifurcated at its base and this should be counted as a single ray. The relative position of dorsal and anal fins on the body with respect to each other and to other body structures is also an important character for identification.

(d) Caudal Fin

Principal caudal fin rays and their supporting bones (hypurals) are the second component to be completely formed after myomeres; so they are much helpful in identification. The secondary caudal fin rays or procurrent caudal fin rays which are supported by epurals on the dorsal side and haemal spines on the ventral side are not important in identification (Fig. 1).

(3) Branchiostegal Rays

The adult branchiostegal ray complement forms early in the larval development; hence the number and relative position of branchiostegal rays are important in identifying the larvae.

B. Morphometric Characters

Morphometric characters are of less importance in identification of larvae as compared to the meristic characters. In some groups measurements of body parts will help to compare shape and proportion of the larval stages with those of juveniles and adults (Fig. 2). This comparison will be useful in relating an unknown larval form to its juvenile and adult. In many other groups the shape and proportion of larvae do not have a useful relation to their adults. In these larvae the body measurements will be helpful to trace the change in shape and proportion during larval development.

The formation of caudal fin has an effect upon measurements of larval length. The principal caudal fin rays and their supporting bones (hypurals) are laid down ventrally under the notochord, a short distance before its tip, and then brought into their terminal position by the upward flexing of the posterior portion of the notochord. The measurement of body length made before flexion from tip of snout to end of notochord is termed as notochord length (NL) (Fig. 3 A, B). After the caudal fin flexion, the measurement from the tip of the snout to the posterior margin of the hypural bone is termed as standard length (SL) (Fig. 3 B, C).

(1) Size and Shape

The relative length of specimens considered at a particular stage of development such as length at hatching, at flexion, at transformation and of juvenile will be useful in identifying the larvae. Size of head compared with the body; size of the eye; length of snout compared to the head; shape and inclination of the gape of the mouth; presence, number, shape and location of teeth; shape of eye and possible presence of eye stalks; presence of barbels on chin and location and size of the otocyst (otolith) assist in identification (Berry and Richards 1973).

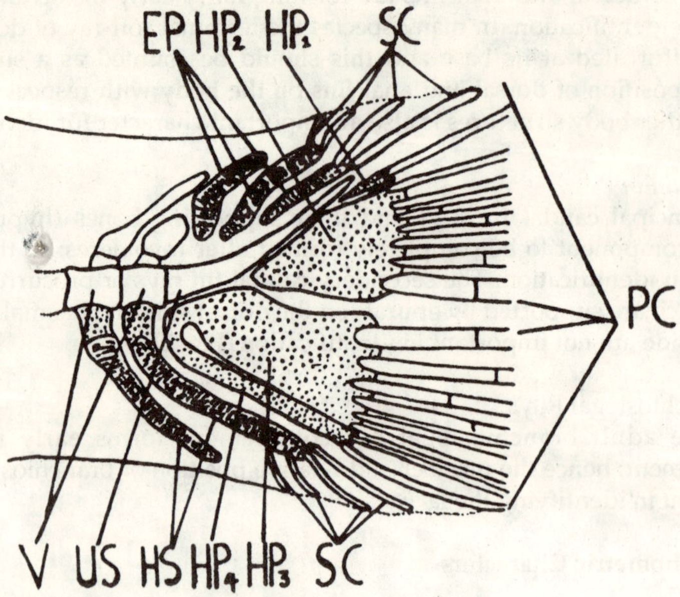

Fig. 1 : Caudal fin skeleton.
EP - Epurals; HS - Haemal spines; SC - Secondary caudal fin rays; V - Vertebrae; HP - Hypurals; PC - Principal caudal fin rays; US - Urostyle

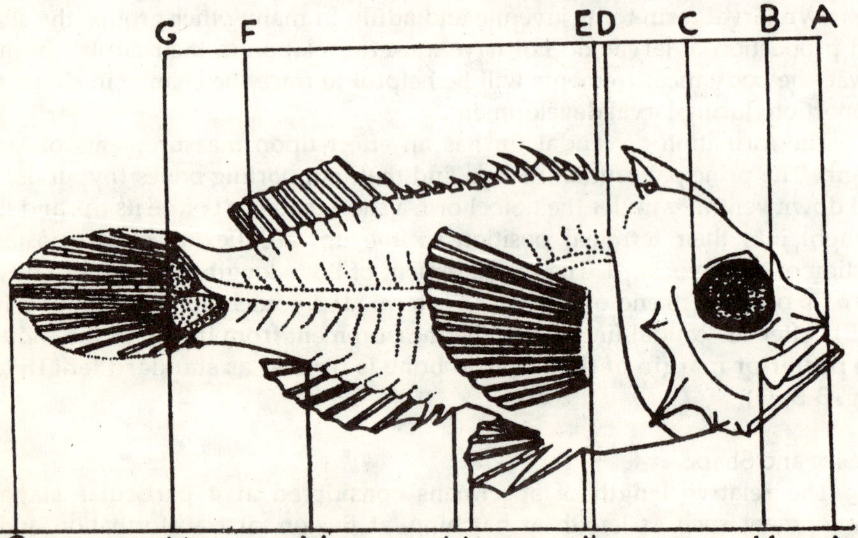

Fig. 2 : A scorpaenid larval fish showing body measurements.
AG - Standard length; AD - Head length; AB - Snout length; BC - Orbital length; AH - Maxilla length; CD - Postorbital length; AE - Predorsal length; AL - Preanal length; AI - Prepectoral length; AJ - Preventral length; EF - Dorsal fin base length; LM - Anal fin base length; JK - Ventral fin length; MN - Caudal peduncle length; NO - Caudal fin length.

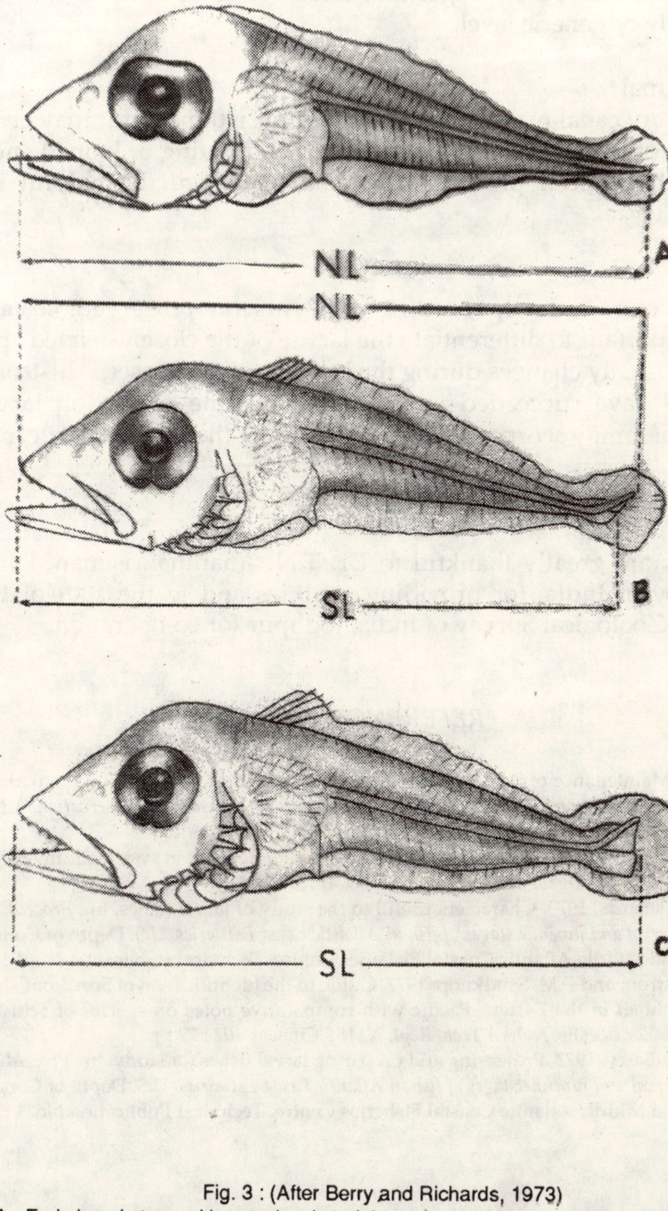

Fig. 3 : (After Berry and Richards, 1973)
A : Early larval stage with notochord straight and caudal bones not formed (Preflexion).
B : Mid larval stage with caudal elements forming and notochord flexing. (Flexion).
C : Late larval stage with caudal elements formed and notochord completely flexed (Postflexion).

(2) Spines

Spines, ridges and crests on the head of the larvae are important for identification, since these structures persist in the juvenile and adult stages. Spines and ridges may occur on the snout, mandible, preopercle, opercle and in the preorbital, suborbital, supraocular, nuchal, pterotic, parietal and posttemporal regions. The location of spines, ridges and crests will be helpful to identify the larvae up to family or generic level.

(3) Alimentary Canal

The alimentary canal may end close behind the isthmus or it may end near the posterior end of the body. It may be straight, undulating or looped and may protrude from the ventral profile. These characters will be helpful in the identification of larvae.

(4) Pigmentation

Black or brown melanophores or pigments on preserved larval fish specimens are important to differentiate the larvae of the closely-related species. Pigment pattern usually changes during the larval growth. Moser, Ahlstrom and Sandknop (1977) have succeeded in differentiating the preflexion larvae of different genera of family scorpaenidae on the basis of the pigment pattern.

ACKNOWLEDGEMENTS

The authors are greatly thankful to Dr. T.N. Ananthakrishnan, Director, Zoological Survey of India, for providing facilities and to the staff of Desert Regional Station, Zoological Survey of India, Jodhpur for co-operation.

REFERENCES

Ahlstrom, E.H. 1976. Maintenance of quality in fish eggs and larvae collected during plankton hauls. In: *Monographs on oceanographic methodology* 4 - *zooplankton fixation and preservation*. UNESCO Press. Edited by H.F. Steedman, pp. 313-319.

Ahlstrom, E.H. and H.G. Moser. 1976. Eggs and larvae of fishes and their role in systematic investigations and in fisheries. *Rev. Trav. Inst. Peches. Marit.*, **40** (3 & 4) : 379-398.

Berry, F.H. and W.J. Richards, 1973. Characters useful to the study of larval fishes. In : *Proceedings of a workshop on egg, larval and juvenile stages of fish in Atlantic Coast Estuaries.* U.S. Deptt. of Commerce, NOAA, NMFS and Middle Atlantic Coastal Fisheries Centre. Technical Publication No. 1 : 37-48.

Moser, H.G., E.H. Ahlstrom and E.M. Sandknop. 1977. Guide to the identification of Scorpionfish larvae (Family Scorpaenidae) in the Eastern Pacific with comparative notes on species of *Sebastes* and *Helicolenus* from other oceans. *NOAA Tech. Rept. NMFS Circular 402* : 77 pp.

Richards, W.J. and F.H. Berry. 1973. Preserving and preparing larval fishes for study. In : *Proceedings of a workshop in egg, larval and juvenile stages of fish in Atlantic Coast Estuaries.* U.S. Deptt. of Commerce, NOAA, NMFS and Middle Atlantic Coastal Fisheries Centre. Technical Publication No. 1 : 12-20.

Rec. Adv. Fish Ecol. Limn. Eco-conserv.
IV : 21-28

5

Observations on the Histomorphology of Pituitary of Grass Carp (*Ctenopharyngodon idella*) and Silver Carp (*Hypophthalmichthys molitrix*)

H.A. KHAN, RADHA C. DAS and P.V.G.K. REDDY

Abstract

The pituitary gland in both grass carp and silver carp is divided into a glandular portion—the adenohypophysis and a nervous portion—the neurohypophysis. The former, as in other teleosts, is divisible into three lobes : rostral pars distalis and proximal pars distalis and pars intermedia, arranged one above the other from anterior to the posterior end. The neurohypophysis ramifies extensively in pars intermedia and comparatively less in other lobes. The neurohypophysial axonal tracts have abundant neurosecretory material. The adenohypophysis in both the above species has eight cell types which are distinguished on the basis of their tinctorial responses. The structural pattern in both the species is observed to be of similar type.

INTRODUCTION

In spite of many works and excellent reviews published on different aspects of pituitary glands of teleostean fishes (Pickford and Atz 1957, Ball and Baker 1969, Holmes and Ball 1974, Kaul and Vollrath 1974, Fontaine and Olivereau 1975), detailed studies on the hypophysis of Asiatic major carps are comparatively few (Sathyanesan 1958, Khan 1962, Lal 1964, Sen 1965, Rai 1966, Rao 1969, Moitra and Sarkar 1976 and Jose and Sathyanesan 1977).

In the present study an attempt has been made to study the histophysiology of pituitary glands of grass carp (*Ctenopharyngodon idella*) and silver carp (*Hypophthalmichthys molitrix*)—the two economically important exotic species which have been introduced in India for aquaculture along with Indian major carps, so as to have a background to study the functional aspects of the adenohypophyseal cells in relation to spawning of these carps.

MATERIALS AND METHODS

Adult silver carp (total length 365-482 mm) and grass carp (total length 350-470 mm) used in the investigations were collected in live condition during their breeding season, i.e. July and August. Immediately after collection, fishes were killed by decapitation. The pituitaries were removed and fixed in aqueous Bouin's fluid containing 5 per cent mercuric chloride. Paraffin sections both in longitudinal and transverse planes were cut at 4-5 μ and stained with:

(a) Alcin blue (AB, pH : 0.2 and 3.0), PAS-Orange G. technique, using acid permanganate oxidation according to Herlant (1960);
(b) Lead haematoxylin (PbH) technique of MacConaill (1947) preceded by PAS;
(c) Cleveland and Wolfe's trichrome, and adopted by Herlant (1965);
(d) Mallory's triple stain;
(e) Heidenhain's Azan method;
(f) Herlant's tetrachrome (1960), as modified by Racadot (1962); and
(g) Aldehyde Fuchsin (AF) technique of Gomori

In order to distinguish different types of cyonophils which undergo histological changes in relation to spawning of both species the following experiments were conducted. In the first group the control fishes did not receive any carp pituitary hormogenate, whereas in the group II, both silver carp and grass carp were injected intramuscularly with aqueous homogenate of alcohol-preserved pituitary, collected from mature donors of Indian major carps at 12-15 mg/kg/body weight of fish split in two doses. Two lots of such injected fishes were collected: 2(a) those which were injected but did not spawn and 2(b) those injected fishes which spawned successfully, pituitary glands collected from these experimental fishes were processed as above.

RESULTS

The pituitary gland in both silver carp and grass carp is a small pear-shaped, whitish, soft body lying on the ventral side of the brain immediately behind the optic chiasma in a concavity situated on the floor of the cranium—the sella turcica. The gland is leptobasic and attached to the infundibulum by a distinct short stalk.

The pituitary gland in both the fishes is divided into a glandular portion—the adenohypophysis and a nervous portion—the neurohypophysis (Fig. A). The former, like other teleosts, is divisible into three lobes, rostral pars distalis (RPD) and proximal pars distalis (PPD) and pars intermedia (PI), arranged one above the other from anterior to the posterior end. The neurohypophysis ramifies extensively in PI and comparatively less in other lobes. The neurohypophysial axonal tracts have abundant neurosecretory material (NSM), stainable with AF, AB and PbH. The adenohypophysis in both the above species has eight cell types which are distinguished on the basis of their tinctorial responses. The structural pattern in both the species is observed to be of similar type.

Rostral Pars Distalis

In both the species investigated, the rostral pars distalis consists of three cell types. The size of different cell types is shown in Table 1.

Fig. 1 : A portion of the RPD (grass carp, *Ctenopharyngodon idella* - control) showing acidophils (long arrow) and cyanophils (small arrow). Herlants tetrachrome X 750.

Fig. 2 : A portion of the PPD (*C. idella* - control), showing acidophils (long arrow) and cyanophils (small arrow). AB-PAS-OG. X 750.

Fig. 3 : Rostral Pars distalis of *C. idella* (control). Upper arrow indicates PAS positive cell and lower arrow indicates PbH positive cell. PAS-PbH. X 750.

Fig. 4 : A portion of the Pars Intermedia of *C. idella* (control). Upper arrow indicates PAS positive cell and lower arrow indicates PbH positive cell. PAS - PbH. X 750.

Fig. 5 : A portion of the RPD of *C. idella* (Treatment 1a). Arrow shows the cyanophils. AB-PAS-OG. X 750.

Fig. 6 : A portion of the RPD of *C. idella* (Treatment 1 b). Arrows indicate the cyanophils. Mallory's triple stain X 750.

Fig. 7 : A portion of PPD of *C. idella* (Treatment 1 a). Arrow (small) indicates the cyanophils which are not degranulated. AB-PAS-OG. X 750.

Fig. 8 : A portion of PPD of (Treatment 1 b). Arrows indicate the vacuolated cyanophilic cells. AB-PAS-OG. X 750.

Fig. 9 : A portion of RPD of *H. molitrix* (control). Long arrow indicates the acidophilic cells and small arrows indicate cyanophilic cells. Cleveland-wolfe's trichrome X 750.

Fig. 10 : Proximal pars distalis of *H. molitrix* (control) showing acidophils (long arrow) and cyanophils (small arrows). Herlants tetrachrome X 750.

Fig. 11 : A portion of RPD of *H. molitrix* showing PbH positive cells (arrow). PAS-PbH. X 750.

Fig. 12 : A portion of Pars Intermedia of *H. molitrix*. Upper arrow indicates PAS positive cells and lower arrow indicates PbH positive cell. PAS-PbH. X 750.

Fig. 13 : A portion of the RPD of *H. molitrix* (Treatment 1 a). Small arrow indicates the cyanophils and long arrow indicates the acidophils. Cleveland-Wolfe's trichrome X 750.

Fig. 14 : A portion of the RPD of *H. molitrix* (Treatment 1 b). Arrow indicates the cyanophils. Mallory's triple stain X 750.

Fig. 15 : A portion of the PPD of *H. molitrix* (Treamtnet 1 a). Arrows indicate cyanophils which are not degranulated. AB-PAS-OG. X 750.

Fig. 16 : A portion of the PPD of *H. molitrix* (Treatment 1 b). Arrows indicate the vacuolated cyanophils. AB-PAS-OG. X 750.

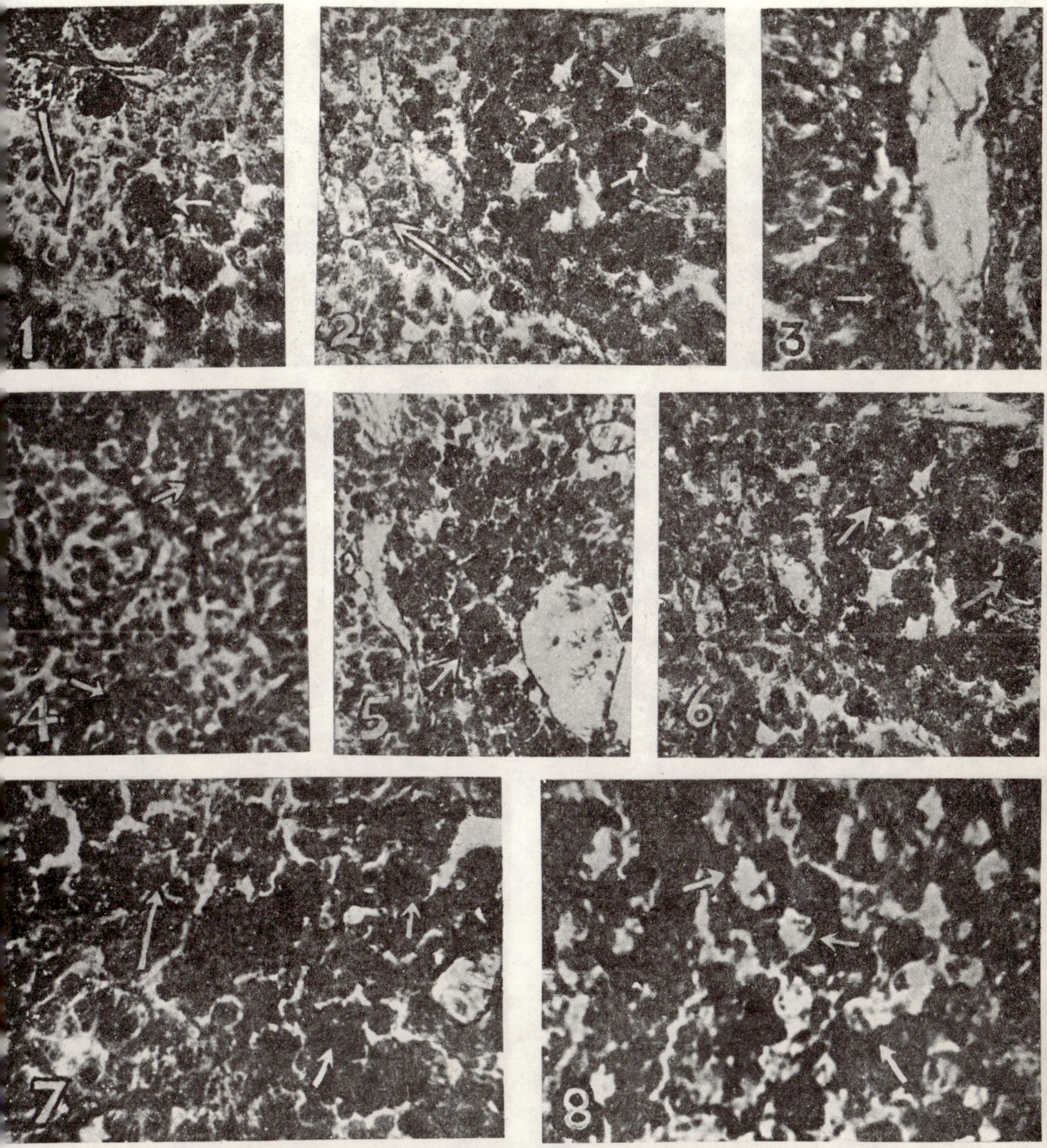

GRASS CARP

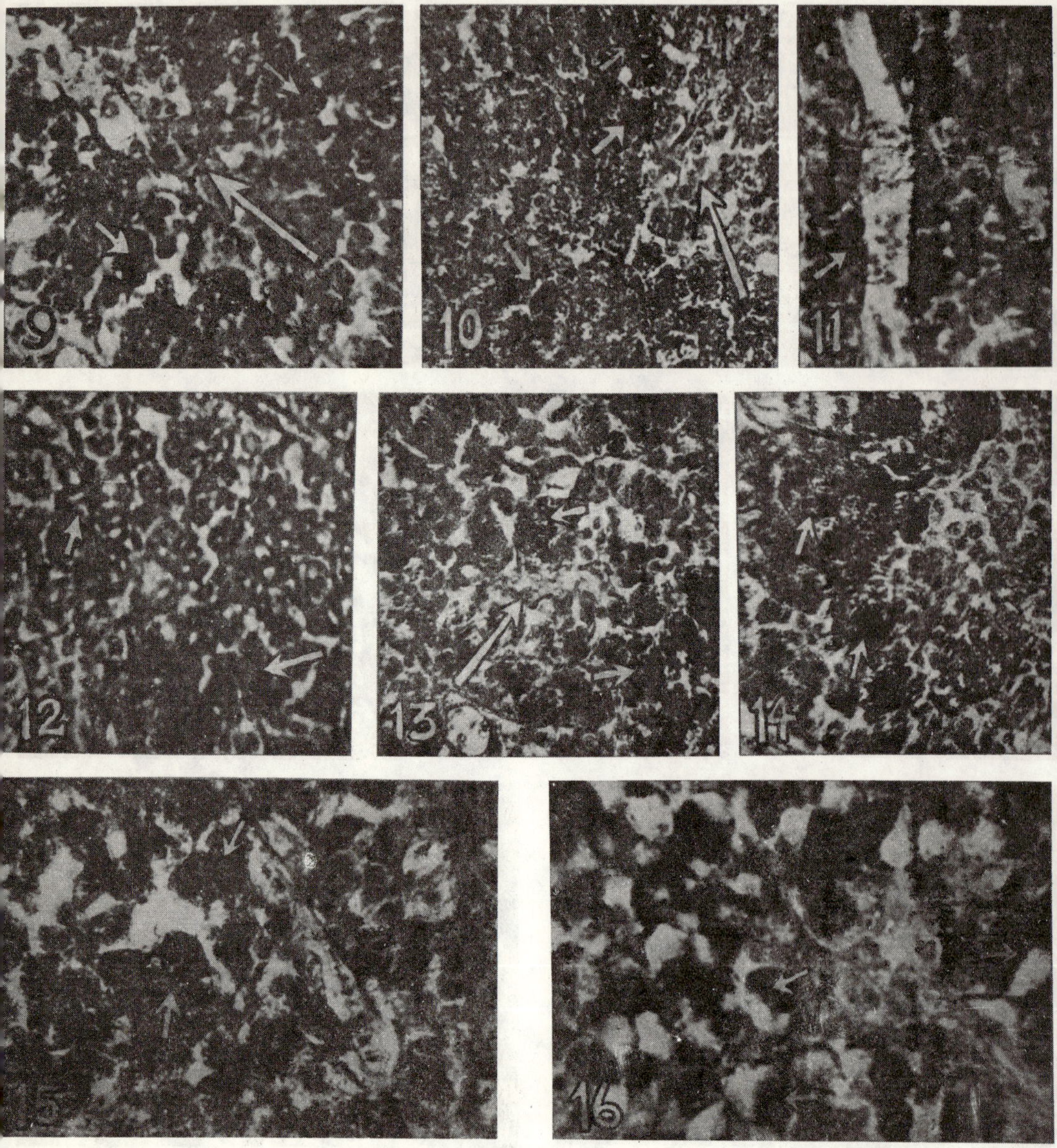

SILVER CARP

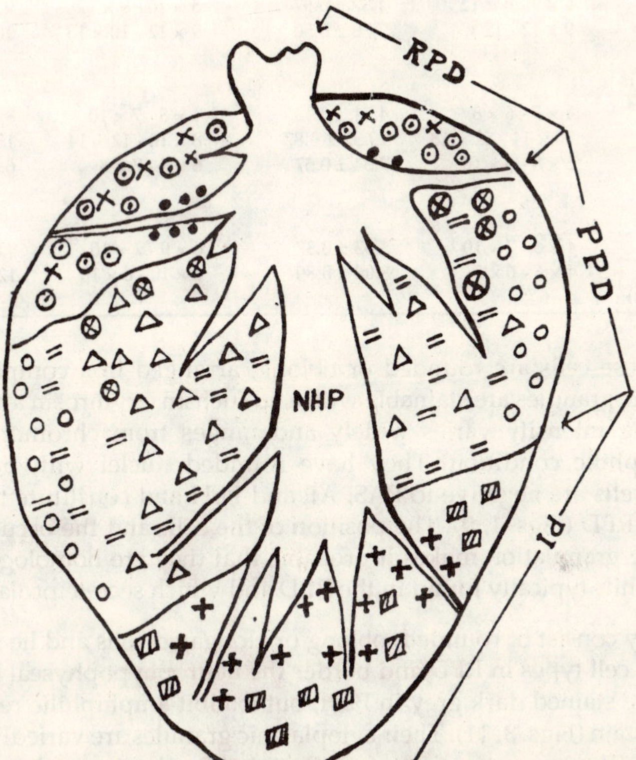

1. ⊙ Type Acidophils
2. ● Type PbH Positive
3. × Type Cyanophil
4. = Type Acidophil
5. Δ Type Cyanophil - I
6. \+ Type PbH Positive
7. ▨ Type PAS Positive
8. O Type chromophobe

Fig. A : Diagrammatic sagittal section of silver carp, *Hypophthalmichthya molitrix* pituitary showing the distribution of different cell types.
NPH - Neurohypophysis, RPD - Rostral Pars Distalis,
PPD - Proximal Pars Distalis, PI - Pars Intermedia.

Table 1 : Size (μ) of Various Adenohypophyseal Cells in Grass Carp and Silver Carp

Various Cell Types	*Grass Carp*		*Silver Carp*	
	Cell size (range in μ)	*Nuclear area (μ^2)*	*Cell size (range in μ)*	*Nuclear area (μ^2)*
Rostral pars distalis				
Type 1	5 × 8 - 6 × 9	4.08 ± 0.24	4 × 8 - 6 × 9	5.77 ± 0.31
Type 2	4 × 9 - 8 × 12	4.32 ± 0.57	5 × 10 - 8 × 12	9.82 ± 0.87
Type 3	9 × 12 - 12 × 4	7.78 ± 0.50	9 × 12 - 10 × 13	10.99 ± 0.52
Proximal pars distalis				
Type 4	4 × 7 - 6 × 8	4.71 ± 0.48	4 × 8 - 7 × 10	8.95 ± 0.30
type 5 (I)	10 × 14 - 12 × 18	19.32 ± 0.87	8 × 10 - 12 × 14	13.67 ± 0.56
Type 5 (II)	6 × 8 - 6 × 10	7.34 ± 0.57	6 × 8 - 7 × 8	6.60 ± 0.49
Pars intermedia				
Type 6	4 × 6 - 7 × 10	3.53 ± 0.37	6 × 9 - 7 × 10	9.82 ± 0.93
Type 7	4 × 6 - 6 × 8	8.09 ± 0.80	4 × 10 - 6 × 10	12.57 ± 0.89
Type 8 (Chromophos)	4 - 6			

Type 1 cells : These cells are rounded or oblong, arranged in a compact mass. Their cytoplasmic granules are stainable with acid fuchsin, erythrosin and orange G. Their staining intensity varies widely and ranges from chromophobic to intense chromophilic condition. They have rounded nuclei with prominent nucleoli. These cells are negative to PAS, AF and PbH and constitute the major cells type of the RPD (Figs. 1, 9). The position of the cells and the occurrence of erythrosinophilic granulation makes it probable that they are homologous with the erythrosinophils typically found in the RPD and which secrete prolactin.

Type 2 cells : They consist of rounded, oblong or elongated cells and lie scattered among the other cell types in RPD and border the neurophypophyseal branches (Fig. A). They are stained dark grey in PbH, but exhibit amphiphilic reaction in Mallory's triple stain (Figs. 3, 11). Their cytoplasmic granules are variedly stained with erythrosin. They are negative of PAS and AB. However, they show a reddish brown hue with AB-PAS-OG, which enables them to be readily differentiated from type I cells present in the RPD.

Type 3 cells : The type 3 cells which are a few in number, are generally round or oval, and are dispersed in the RPD (Figs. 1, 9). These cells are cyanophils and are PAS + AF + AB + and aniline blue positive. These cells are found to be heavily granulated and did not vacuolate even in the spent fishes.

Proximal Pars Distalis

The proximal pars distalis is the largest region in the pituitary of both species, and consists of acidophils, cyanophils and chromophobes.

Type 4 cells : These are the classical acidophils which are distributed in patches throughout the PPD without any specific pattern of their arrangement (Figs. 2, 10). These cells may be angular, slightly elongated or rounded in shape. Their nuclei may be oval or rounded. The cytoplasm is densely granulated. They are stained with acid fuchsin, orange G and erythrosin and do not respond to AB, PAS, AF, PbH or aniline blue.

Type 5 cells: The type 5 cells are the characteristic cyanophils present in large masses in the PPD region of the pituitaries of both the species and are stained with PAS, AF, AB and aniline blue; some of these cells were bigger in size and contained globules.

The type 5 cells only responded differently to various treatments. In the untreated control fishes, these cells are heavily granulated (Figs. 2, 10), whereas in experimental groups, which received hormonal treatment and spawned, most of these cells were totally degranulated and distinctly vacuolated (Figs. 8, 16). However, in the other group of experimental fishes which were administered with carp pituitary hormogenate but did not spawn, these cells showed very little of degranulation and vacuolation. Thus these histological changes occurring in Type 5 cells correlated with spawning represent their nature as gonadotraphs and these cells may be designated as Cyanophils I of PPD. The other type of cyanophils, termed as Cyanophil II, consisted of cells smaller in size than Cyanophils I and laid interspersed in patches, largely resembling the type 3 cells of RPD in their staining reaction. These cyanophil II cells did not show much degranulation either in untreated control or in treated fishes of both the species. However, to differentiate Cyanophils II from Cyanophils I needs experimental verification.

Chromophobes are also a few in number and lie scattered in between the cynophils and acidophils. They are irregular in shape and consist of almost centric nuclei. Some times they vary in their size from 4×6 – 6 × 8 μ in both species.

Pars Intermedia

This lobe has two distinct types of cells in both the species studied.

Type 6 cells : They are positive to PbH which are similar to the colour reaction of type 2 cells in the RPD (Figs. 4, 12). Several of them are arranged bordering the neurohypophyseal branches of PI. These cells show amphiphilic reaction to Mallory's triple stain and has a strong affinity for erythrosin.

Type 7 cells : These cells are lightly PAS positive (Figs. 4, 12) which do not respond to AB, AF or aniline blue.

Some chromophobes are observed in PI region. The chromophobes in RDP, PPD and Pi constitute the eight cell type and may be considered as immature or degranulated chromophils.

DISCUSSION

Eight histologically distinct types of cells can be identified in the pituitary of silver carp and grass carp, based on their tinctorial response, obtained for the various staining techniques used by several investigators and are comparable to that of pituitaries of other carps.

The cells of rostral pars distalis (RPD) in both the species studied in the present investigation are typical in that they consist of abundant acidophils, amphiphils which are deeply stained by PbH, and a few cyanophils. The colour reaction of acidophils varies from chromophobic to intense chromophilic condition. Similar observations have been made in *Cyprinus carpio* and *Carassius auratus* (Scruggs 1951), and *Labeo rohita* (Moitra and Sarkar 1976 and Jose and Sathyanesan 1977).

The prolactin cells of RPD are reported to be either chromophobic or are packed with azocarmin, acid fuchsin or erythrosin-staining granules. In silver carp and grass carp the type 1 cells, which are stainable with acid fuchsin and erythrosin, depending on the technique employed, form the major component of RPD, exhibit different staining intensities and some being even chromophobic. Judging from staining properties the Type 1 cells may correspond to the presumptive prolactin cells. This can be partly concluded from the fact that in teleosts the acidophilic prolactin cells are restricted to this lobe (Ball and Baker 1969 and Kaul and Vollrath 1974).

The PbH positive cells of grass carp and silver carp based on their staining properties can be categorized as ACTH cells and have the same localization, bordering the neurohypophysis, as is common in other teleosts (Olivereau 1964, 1970, Ball and Baker 1969, Oguri 1971 and Kaul and Vollrath 1974). In *Labeo rohita*, these cells have been reported lying among the other component cells of the RPD and also bordering the neurohypophysial branches (Jose and Sathyanesan 1977).

In grass carp and silver carp three types of cyanophils have been identified in PPD with the aid of recent staining techniques. These are Type 3 cells in RPD and Type 5 Cyanophils I and Type 5 Cyanophils II in PPD. But in silver carp Yang Ali-sheng (1966) described 3 types of cyanophils in the PPD. Cyanophils represent thyrotrophs (TSH cells) and gonadotrophs (GTH cells) which resemble in their staining reaction. Attempts to distinguish them purely on the basis of their tinctorial affinities have been inconclusive. The distributional pattern and site of TSH cells also seem to vary among teleosts. In *L. rohita* Khan (1962) reported the absence of cyanophils in RPD, whereas their presence has been reported in the same species by Moitra and Sarker (1976) and Jose and Sathyanesan (1977). Moitra and Sarkar (1976) observed the cyanophils situated towards the periphery of RPD to differ both in shape and size from those located adjacent to PPD. Contrary to that, Jose and Sathyanesan (1977) reported the cyanophils cells located both in the area comprising of anteroventral RPD and anterodorsal PPD to be comparable to TSH cells, based on their staining reaction, shape and distributional pattern. The results obtained in the present study indicate that cyanophils (cell type 3) located in the RPD of both silver carp and grass carp and cyanophils II (Type 5 cells) in the PPD are involved in TSH production, as both cell types exhibited very little of degranulation during the spawning phase of both untreated control and treated fishes (under treatments 2a and b).

Type 5 cells of PPD (Cyanophils I) have been identified as a single category of gonadotrophs which are heavily granulated in the untreated control fishes, but they undergo active degranulation and vacuolization and lose their contents in those treated fishes which spawned and became spent (under treatment 2 b). The mean nuclear area of degranulated cyanophils under treatment 2b were $8.93 \pm 0.62\ \mu^2$ and $6.44 \pm 0.53\ \mu^2$ in grass carp and silver carp respectively and were highly significant when compared to those in the untreated species (Table 1). In grass carp only one type of gonadotrophs in the PPD containing two kinds of granules differing in size and tinctorial reaction have been reported (Anon. 1978, quoted by Pritchard 1980).

The acidophils of the PPD which are organgeophils have been reported to synthesise somatotropin (Ball and Baker 1969, Sage and Bern 1971). The location of STH cell varies in different species. In *Labeo rohita* Moitra and Sarkar (1976) distinguished two types of acidophils—one erythrosinophilic and the other orangeophils as somatotrophs and the erythrosinophils as the cells producing prolactin. Contrary to the findings of these workers, Jose and Sathyanesan (1977) reported only one type of acidophils in the PPD of *L. rohita*, which they suggested to be comparable to STH cells based on their staining reaction. However in silver carp and grass carp since there is only one type of acidophils (Type 4 cells) observed in the PPD without any specific pattern of arrangement, being distributed in patches throughout the PPD, there is no doubt that it is this cell which is involved in the production of growth hormone (Kaul and Vollrath 1974).

Two types of cells have been distinguished in the pars intermedia of teleosts which may be acidophilic, basophilic or amphiphilic or chromophobic. In silver carp and grass carp, some of the cells which are acidophilic to trichrome and tetrachrome techniques are apparently PbH positive and those that are amphiphilic are found to be highly PAS positive.

SUMMARY

To summarise, the results obtained in this study indicate that 8 cell types are distinguishable among the adenohypophysial cells of silver carp and grass carp, using several combinations of staining so the cell types 1, 2, 3 and 5 (type II), 4, 5 (type I) of both species are comparable tentatively to the eta (prolactin), epsilon (corticotrophs), delta (thyrotrophs), alpha (somatotrophs) and beta and gamma (gonadotrophs) cells, described in the published works on teleostean hypophysis and need further experimental proof. However, more precise morphological and physiological observations are needed to elucidate this field of functional cytology.

ACKNOWLEDGEMENTS

The authors express their thanks to Dr. V.G. Jhingran, former Director and Dr. A.V. Natarajan, officiating Director for encouragement. They are also indebted to Dr. V.R.P. Sinha, National Project Director and Head of FARTC, Bhubaneswar for providing the laboratory facilities, guidance and helpful suggestions in the preparation of the manuscript, to Shri R.D. Chakraborty, Head of Pond Culture Division and to Dr. S.B. Singh, Project Coordinator, All India Coordinated Project on Composite Fish Culture and Fish Seed Production for their valuable suggestions and encouragement during the course of this investigation. Thanks are also due to Prof. S.K. Moitra of Burdwan University and Dr. A.G. Sathyanesan of Surgical Research Laboratory, Benaras Hindu University for kindly providing the authors with some stains.

REFERENCES

Ball, J.N. and B.I. Baker 1969, The pituitary gland: anatomy and histophysiology. In: *Fish Physiology*. Eds. W.S. Hoar and D.J. Randall, Vol.2, Academic Press, New York : 1-93.

Fontaine, M. and M. Oblivereau, 1975, *Amer. Zool.*, **15** : 16-79.

Herlant, M., 1956, *Arch. Biol. Liege*, **67** : 89-180.

Herlant, M., 1960, *Bull. microse. appl. Paris*, **10** : 37-44.

Holmes, R.L. and J.N. Ball, 1974, *The pituitary gland*. Cambridge University Press, Cambridge : 397 p.

Jose, T.M. and A.G. Sathyanesan, 1977, *Anat. Anz.*, **142** : 410-423.

Kaul, S. and L. Vollrath, 1974, *Cell Tiss. Res.*, **154** : 211-230.

Khan, H.A., 1962, *Proc. Nat. Acad. Sci. India*, **32**(B) : 101-114.

Lal, B., 1964, *Proc. Indian Acad. Sci.*, **59**(B) : 297-317.

Mac Conaill, M.A., 1947, *J. Anat. London*, **81** : 371-372.

Moitra, S.K. and S.K. Sarkar, 1976, *Ant. Anz.*, **139** : 421-441.

Oguri, M., 1971, *Bull. Jap. Soc. Sci. Fish*, **37** : 577-584.

Olivereau, M., 1962, *C.R. Acad. Sci. Paris*, **255** : 2007-2009.

Olivereau, M., 1964, *Z. Zellforsch. u. mikrosk. Anat. Berlin(W)*, **63** : 496-505.

Olivereau, M., 1957, *Acat. Zool.*, **51** : 229-249.

Pickford, G.E. and J.W. Atz., 1957, *The physiology of the pituitary gland of fishes*. Zoological Society, New York.

Racadot, J., 1962, *Bull. Microse, appl. Paris*, **12**: 16-20.

Rai, B.P., 1967, *Anat. Anz.*, **120** : 360-374.

Rao, P.D.P., 1969, *Acat. anat.*, **73** : 281-303.

Rao, P.D.P., 1971, *Zool. Anz.*, **184** (5/6) : 335-348.

Sage, M. and H.A. Bern, 1971, *Internat. Rev. Cytol. New York*, **31** : 339-376.

Sathyanesan, A.G., 1958, *Indian J. Vet. Sci. Anim Husband.*, **28** : 13-20.

Sen, P.R., 1972, *J. Inland Fish. Soc. India*, **4** : 20-25.

Yang Ali-Sheng, 1966, *Acta Anat. Sinica*, **9**: 100-107.

Anon. 1978, *Sci. Sin*, **21** (3) : 397-400 (quoted by Pritchard, G.I. (1980): *Fisheries and Aquaculture in the Peoples' Republic of China*, Ottawa, Ont. IRDC : 32 p.)

Rec. Adv. Fish Ecol. Limn. Eco-conserv.
IV : 29-32

6

Nucleolar Extrusions in the Oocytes of Nepalese Snow Trout, *Schizothorax plagiostomus* (Heckel)

TEJ KUMAR SHRESTHA

Abstract

The phenomenon of nucleolar extrusion in the oocytes of *Schizothorax plagiostomus* has been critically studied. Nucleoli pass out as entire bodies without disrupting nuclear envelope. The pocket formation of nuclear envelope does not necessarily precede the extrusions. The extruded nucleoli appear to have some role in the process of vitellogenesis. Nucleoli in some oocytes are seen arranged in double rows or rings. The duplex arrangement of nucleoli in occytes appear to be interesting.

INTRODUCTION

Several important contributions on nucleolar activity of fish oocytes are available, viz., Scharff (1888), Wallace (1903), Eggert (1929), Nath and Nangia (1931), Singh and Boyle (1935), Chaudhry (1951), Yamamoto (1956), Malone and Hisaoka (1963), Guraya, Kaur and Saxena (1975), Kapoor (1977) and Rita Kumari and Nair (1979). So far, only a few reports are available on the nucleolar extrusion in the oocytes of hill-stream fishes (Khanna and Pant 1967, Bisht and Joshi 1975 and Shrestha 1980ab). In the present paper, nucleolar extrusion in the oocytes of hill-stream fish, *Schizothorax plagiostomus*, have been described.

MATERIALS AND METHODS

Monthly samples of *S. plagiostomus* were collected from Panauti Roshi river at an altitude of 1500 m. The small pieces of the ovary were washed thoroughly in normal saline and fixed in Bouin's and Smith's fixatives respectively. The pieces, of the ovary contained in Smith's fixatives were kept in dark for 48 hr and were washed in 2 per cent formalin. In both cases section were cut at 5 micra after embedding in Paraffin wax. Serial sections were cut at 5-6 μ and stained in Heidenhain's iron haematoxylin, Mallory's triple stain and mercuric bromophenol blue.

OBSERVATIONS

The nucleus of the early maturing oocyte, termed as germinal vesicle, is the most important organelle. The oocyte in the chromatin nucleolus stage has large biscuit shaped nucleus which covers the major portion of the ooplasm (Fig. 1). In this stage the oocyte nucleus shows reticular network of chromatin particles which are seen scattered at random in the nucleoplasm. Later on, when the oocyte reaches peri-nucleolus stage, the nucleus assumes spherical shape. The lampbrush chromosomes characteristic of the teleostean and amphibian oocytes appear in this stage. In the primary yolk stage these chromosomes disappear in the nucleoplasm. In the early oocyte the nuclear membrane is very thin and conspicuous. As oocytes advance towards maturity they become gradually inconspicuous. In the early stage of the nucleolar extrusion, only one or two nucleoli of varying size are visible. Later on as oocytes grow the number of nucleoli increases while their size decreases. This suggests formation of extra-nucleoli by division and fragmentation of the original nucleolus (Figs. 5 and 6).

The process of nucleolar extrusion in the oocytes of *S. plagiostomus* can be divided into five stages. During the first phase of nucleolar extrusion nucleoli of different sizes can be observed lying at the centre of the nucleus (Fig 1). In the second phase the extruded nucleoli start their movement towards the periphery, abutting closely the nuclear membrane (Figs. 2, 3 and 4). In the third phase the nucleolar membrane is thrown into wavy projections (Figs. 7 and 8) so as to form nucleolar pockets. During the fourth stage of nucleolar extrusion individual nucleoli fuse with each other and the resulting mass comes to lie near the pockets of the nuclear membrane and then scuttle out into ooplasm as solid entity, so that some extruded nucleoli are seen outside the nucleolar membrane (Figs. 6 and 8). In the fifth stage the extruded nucleoli break into small granules and get mixed up into the ground cytoplasm and form proteid yolk. The proteid yolk stains deep blue with mercuric bromophenol blue which is rich in protein and RNA. In this phase some nucleoli are seen passing through nucleolar membrane completely, while others lie partly inside and partly outside of the nuclear membrane (Fig. 3). In all pre-vitellogenic oocytes contraction of the nucleoplasmic material is observable during fixation and dehydration process. But such contraction does not appear to occur in fully mature oocyte. In some pre-vitellogenic oocyte, nuleoli are seen arranged in two distinct circular rings (Fig 9 and 10). The outer ring of the nucleoli lie in the extranuclear zone, i.e. in the vicintiy of the nuclear envelope, and the inner ring lies at the intra-nucleolar zone, i.e. at the core area of the nucleoplasm. Firstly, the nucleoli arranged in the extra-nucleolar zone appear to take part in the emission process and then followed by the nucleoli arranged in the intranucleolar zone. The existence of such a double or duplex ring of nucleoli in some oocytes of the this fish is very peculiar and needs further investigation.

DISCUSSION

There are several views regarding the origin, extrusion and function of nucleoli in teleosts. According to one view a single nucleolus of oogonium gives rise to a number of nucleoli by fragmentation and division. (Chaudhary 1951, 1952; Rai 1967; Shrestha and Khanna 1980). Contrary to this, Yamamoto (1956a), Bara (1960), Khanna and Sanwal (1971), Rita Kumari and Nair (1979) did not observe any fragmentation in the fishes examined by them. According to their

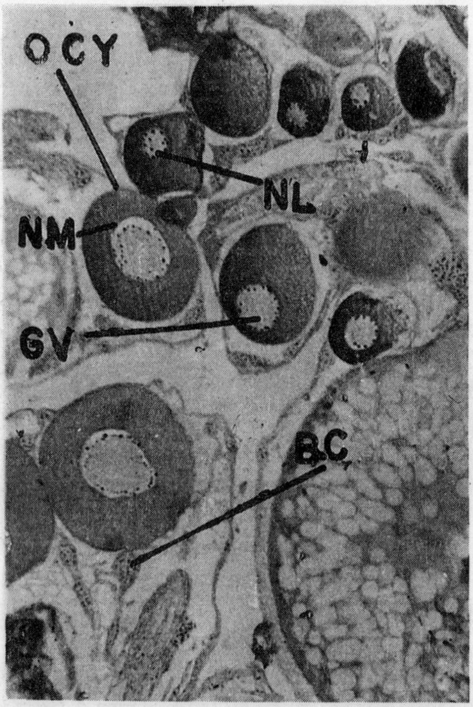

Fig.1 : Pre-vitellogenic oocyte showing movement of nucleoli towards the periphery. Also note contracted karyoplasm in the germinal vesicle X 150.

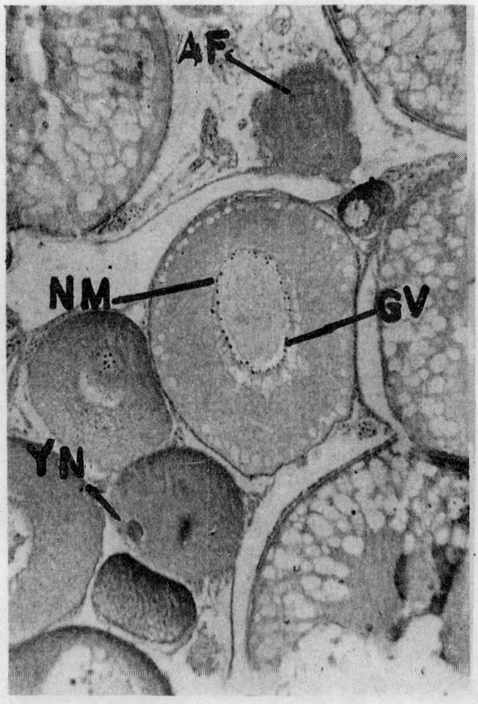

Fig.2 : Germinal vesicle of the oocyte at yolk vesicle stage. Note scattered nucleoli around nucleoplasm. X 170.

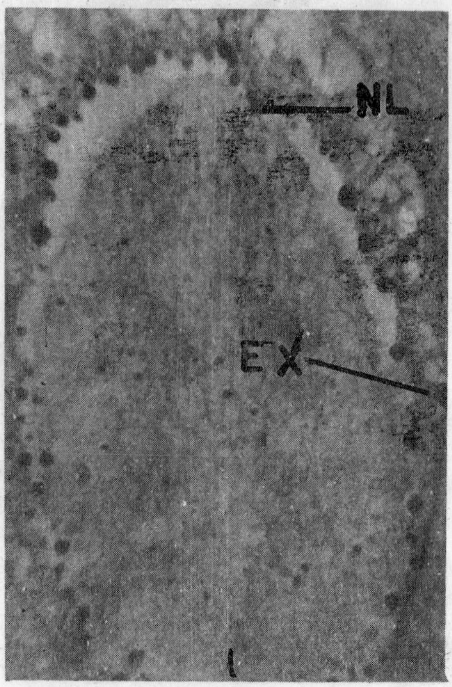

Fig.3 : Germinal vesicle of the oocyte at the beginning of yolk formation. Note progressive nucleolar emission. Some nucleoli are seen partly inside and partly outside of nucleolar envelope X 400.

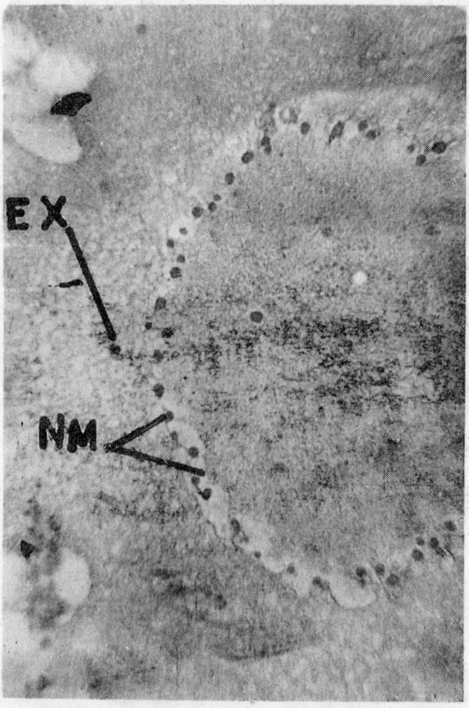

Fig.4 : Germinal vesicle of oocyte at later stage of yolk formation. Note undulated of nuclear membrane with nucleoli against nuclear envelope X 400.

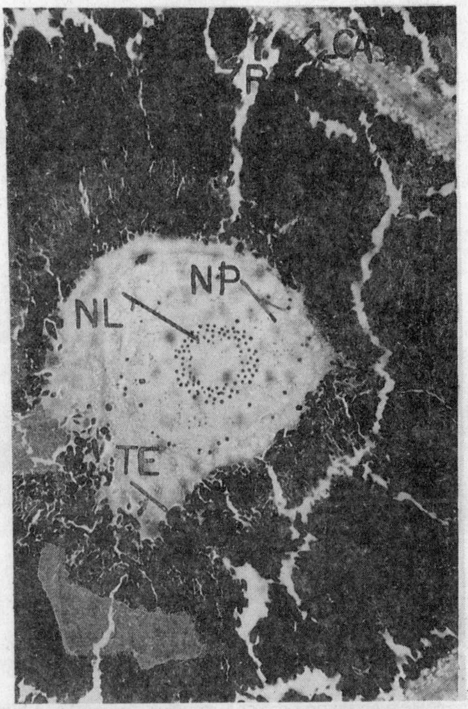

Fig.5 : Nucleolar extrusion in fully mature oocyte. Note reduction of nucleolar diamotor and concentration of nucleoli at the core area of nucleoplasm X 150.

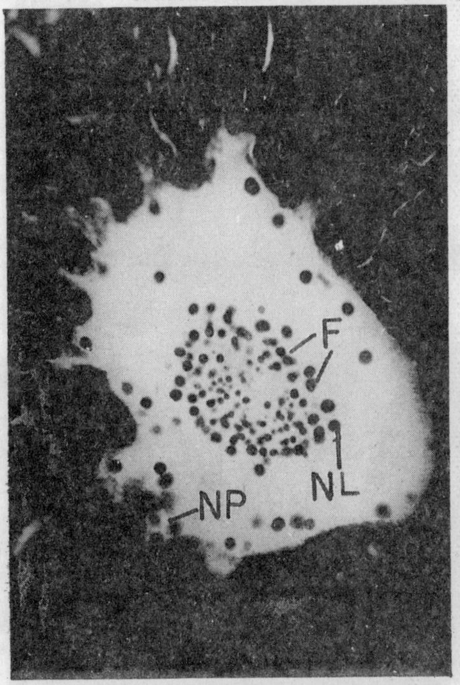

Fig.6

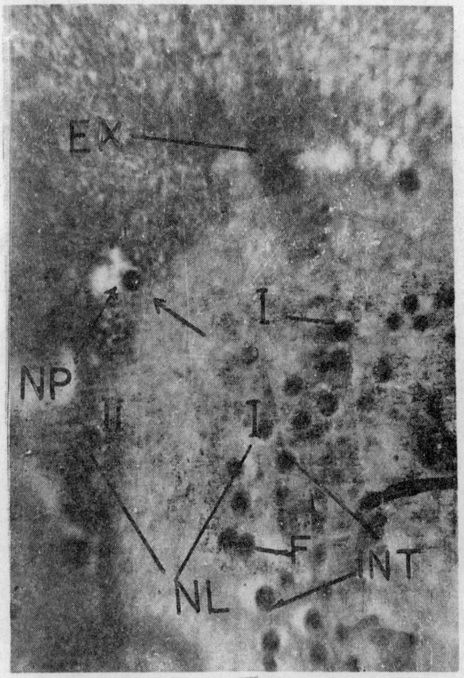

Fig.7

Fig.6 : Showing nucleolar extrusion in nearly vitellogenic oocyte. Note concentration, dispersion and fusion of nucleoli X 150.

Fig.7 : Showing pocket formation in extra-nuclear zone. Nucleoli are pinched off in nucleolar pockets (arrow) X 400.

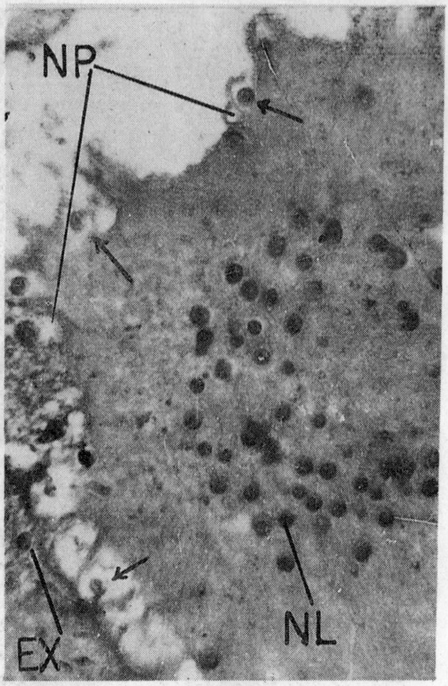

Fig.8 : Part of vitellogenic follicle showing pocket formation and extensive fusion of nucleoli (F) X 400.

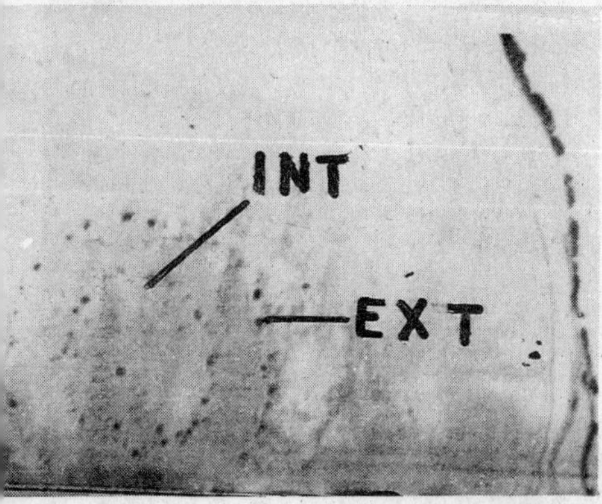

Fig.9

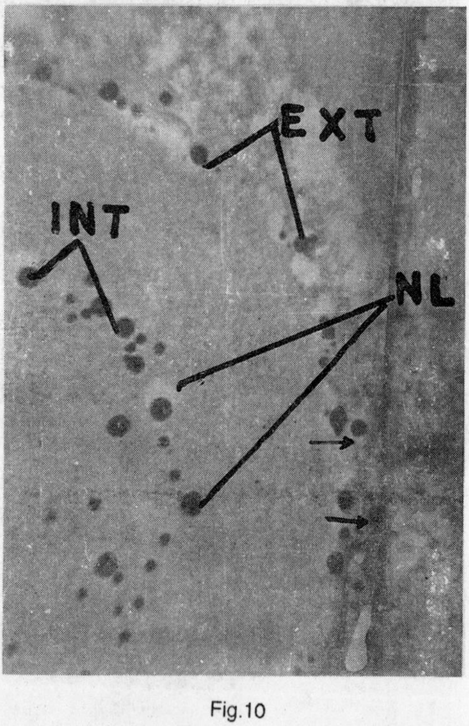

Fig.10

Fig.9 : Pre-vitellogenic follicle of late perinucleolus stage showing nucleoli arranged in intra-nuclear and extra nuclear ring X 160.

Fig.10 : Pre-vitellogenic follicle showing duplex arrangement of nucleoli X 400.

views, peripheral nucleoli are formed by the fusion of minute Feulgen-positive particles present in peripheral nucleoplasm (Bara 1960). According to modern view, the nucleoli originate from certain heterochromatic region of chromosome known as nucleolus organizer (Vincent 1955, Arndt 1966, McGregor 1972). Recently, Chouinard (1963) distinguished a non-heterochromatic region of chromosome from nucleolar organising region, which is also instrumental for the formation of extranucleoli.

In *S. plagiostomus* many nucleoli of various sizes are seen in the late perinucleolus stage. The number of nucleoli goes on increasing as oocyte advances towards maturity and their size gets reduced. When the oocyte reaches migratory nucleolus stage, nucleoli appear as tiny particles. Such progressive reduction in size and accompanied by increase in number clearly suggest that nucleoli arise by division and fragmentation of the original nucleolus.

Extrusion of nuclei into ooplasm has been a subject of controversy. Wallace (1904) and Nath and Nangia (1931) are few important workers who hold view that the nuclear membrane does not allow any transmission of nucleoli through it. Another school of thought is that there is definite extrusion of nucleoli. A few pioneer investigators supporting this view are Narain (1928), Malhotra (1963), Khanna and Pant (1967), Dixit and Agrawala (1974), Kapoor (1977). The study of *S. plagiostomus* does show histologically the stages where the nucleoli appear to be extruded out of the nuclear envelope, thus lending support to the second school of thought.

Nuclear pockets as observed in the oocytes of *S. plagiostomus* has been reported by some authors (Eggert 1929 and Chaudhry 1951). But the elongating and detaching nuclear pockets each with nucleoli are not observed in this fish.

Several worker report that nuclear extrusions do not play a visible role in yolk formation (Eggert 1929 and Guraya et al. 1975). On the contrary many workers suggested extrusion of nucleoli associated with yolk formation (Malone and Hisoaka 1963, Malhotra 1963, Kapoor 1977). Some outhors like Yamamoto (1956) and Bara (1960) suggested that nucleolar extrusions are associated with cell metabolism. Similarly, Raven (1961) postulated the fact that nucleolar emission contributes ribosomal need of rapidly growing previtellogenetic oocyte. Studies on *S. plagiostomus* reveal that RNA and protein-rich nucleolar material on extrusion get distributed in cytoplasm as to form proteid yolk. This observation is in conformity with those of Saxena and Choudhry (1977) in *Colisa fasciatus*.

ABBREVIATIONS

AF. Atretic follicle.
BC. Blood cells
CA. Cortical alveoli
EX. Extruded Nucleoli
F. Nucleoli undergoing active fusion
GV. Germinal vesicle
I. Intranuclear nucleoli
II. Extra-nuclear nucleoli
NL. Nucleoli
NM. Nuclear membrane

NP. Nuclear pockets
TE. Totally extruded nucleoli
YN. Yolk nucleus
ZR. Zona radiata

REFERENCES

Arndt, E.A. 1960. Die Aufgaben des kerns wahrend der oogenese der Teleosten Z. *Zellforsch*, **51** : 356.

Bara, G. 1960. Histotological and cytological changes in the ovary of the mackeral *Scomber scomber* L. during annual cycle. *Rev. Fac. Sci. Univ. Istanbul*, **15** : 49-91.

Bisht, J.S. and M.L. Joshi, 1975. Seasonal histological changes in the ovaries of a mountain stream trout *Schizothorax richardsonii* (Gray & Hard.) *Acta. anta.*, **93** : 512-525.

Chaudry, H.S. 1951 Nucleolar activity in the oocytes of some marine teleosten fishes. *J. Roy. Micr. Soc.*, **71** : 87-93.

Chouinard, L.A. Sites of formation of extra nucleoli during early oocyte growth in freshwater toleost *Salvelinus fontinalis* (Mitchill). *Canad. J. Zool.*, **41**: 997-1010 (1963).

Dixit, R.K. and Agrawala, N. Studies on the developmental rhythm in oocyte of *Puntius sophore* (Ham). *Acta. Anat.*, **96** : 133-144 (1974).

Eggert, B. Entwichlung und Bau der Eier von *Salarius flavoumbrinus* Rupp. *Zool. Anz.*, **83** : 241-253 (1929).

Guraya, S.S. A comparative histochemical study of fish (*Channa marulius*) and amphibian (*Bufo* and *Stomaticus*) oogenesis. *Z. Zellforsch.* **65** : 662-700 (1965).

Guraya, S.S, Kaur. R. and Saxena, P.K. Morphology and ovarian changes during reproductive cycle of fish *Mystus tengra* (Ham). *Acta. anat.* **91** : 222-266 (1975).

Hisoaka, K.K. and Firlt, C.F. The localization of nucleic acids during oogenesis in zebra fish. *Amer. J. Anat.*, **110** : 203-216 (1962).

Kapoor, C.P. Nucleolar extrusion in the oocytes of *Puntius ticto*. (Ham) *Zool. Beitr.*, **23**(2) : 221-226 (1977).

Khanna, S.S. and Pant, M.C. Seasonal changes in the ovary of a sisorid cat fish *Glyptosternum pectinopterum*. *Copeia*, **83** : 88 (1967).

Khanna, S.S. and Sanwal, R. Cyclic changes in the ovaries of a freshwater teleost, *Channa gachua*. *Zool. Beitrage*, **17** : 312-326 (1971).

McGregor, H.C. Nucleoli and genes in oogenesis. *Biol. Rev.*, **47** : 177-219 (1972).

Malhotra, Y.R. On the nucleolar extrusion in the developing oocytes of a Kashmir fish, *Schizothorax niger* (Heckel). *Ichthyologica.*, **II** : 57-60 (1963).

Malone, T.E. and Hisaoka, K.K. A histochemical study of the formation of deutoplasmic components in developing oocytes of zebra fish. *J. Morph*, : 61-75 (1963).

Narain, D. Cytoplasmic inclusion in oogenesis of *Saccobranchus fossilis*, *Clarias batrachus* and *Anabas scandens*. *Z. Zellforsch.*, **26** : 623-640 (1937).

Nath, V. and Nanigia, M.D. A demonstration of the vacuole and golgi apparatus as independent Cytoplasmic components in fresh egg of teleostean fishes. *J. Morph. Physiol.*, **42** : 277-308 (1931).

Rai, B.P. Cyclic changes in the ovaries of *Tor. (Barbus) tor* (Ham). *Acta. Zoologica.*, **47** : 289-309 (1967).

Raven, C.P. *Oogenesis, the storage of developmental information*, Pergamon Press, London, (1961).

Rita Kumari, S.D. and Nair, N.B. Oogenesis in tropical loach *Lepidocephalus thermalis* (Cuv and Val). *Proc. Indian. Acad. Sci.*, **88** : 45-54 (1979).

Saxena, O.P. and Choudhry, V. Nucleolar extrusion in *Colisa fasciatus*. *Proc. Bull. All India seminar on Ichthyology*, Gaziabad (1977).

Singh, B.N. and Boyle, W. The Vitellegenesis of *Gasterosteus aculeatus* (the stickle back) investigated by ultracentrifuge. *Quart, J. Micr. Sci.*, **81** : 83-106 (1936).

Scharff, R. Intra-ovarian eggs of osseous fishes. *Quart. J. micr. Sci.* **81** : 83-106 (1888).

Shrestha, T.K. Maturation of the oocyte and seasonal changes in the ovary of a hill-stream teleost, *Garra gotyla* (Gray). In press.

Shrestha, T.K. Rhythm of oocyte maturation and reproductive cycling in female stone loach *Noemacheilus beavani* (Gunther). *Bull. Acad. Ichthyology. Modinagar.* (In Press.)

Stromastsen, F.A. The development of gonads in the gold fish *Carassius aurastus* L. *Uni. Iowa. Study in Natural History*, **13** : 1-45 (1931).

Wallace, W. Observation on ova and follicles of teleostean and elasmobranch fish. *Quart. J. Micr. Sci.* **47** : 161-213 (1904).

Rec. Adv. Fish Ecol. Limn. Eco-conserv.
IV : 33-36

7

Study of the Ribonucleic Acid in Five Carp Eggs

SUBIR KUMAR MUKHOPADHYAY, CHITTARANJAN SAHU and ASOKE BOSE

The fish egg cytoplasm, like that of other animals, contains, among other constituents, appreciable amount of ribonucleic acid (Brown and Littna 1964, 1966) synthesized in the oocyte nucleus (Smith 1957, Ficq 1955, Gall 1963, Balinsky 1970). This RNA helps the post-fertilisation development of the egg through early gastrula, synthesizing the so-called house-keeping enzymes and other protein characteristics of the species (Crippa *et al.* 1967). As RNA in the egg is mostly transcribed by DNA, its qualitative survey in the eggs of five carps—*Cirrhina mrigala* (Ham.), *Labeo rohita* (Ham.), *Cyprinus carpio* (Linn.), *Hypothalmicthys molitrix* (Valen.), and *Ctenopharyngodon idella* (Valen.) has been made. This may help to establish another parameter in the biochemcial and genetical taxonomy (Simpson 1961, Tsuyuki and Roberts 1966, Sackin 1969) and may thus help to select suitable species for hybridization, which is a common modern practice for improvement of fish stock.

MATERIALS AND METHODS

RNA from the mature eggs of a species of the fish studied was extracted and electrophoresed according to the method of Zschcile and Murray (1963) in 0.55 M sodium chloride and precipitated with 4 volumes of 95 per cent ethanol. This was stored at -10°C for 24 hours, when flakes of RNA become deposited. The sample was purified with chloroform and octanol (4:1) and by repeated precipitations with ethanol. The purity of the sample was tested in a MOM U-V Spectrophotometer at 260 and 280 mm.

ELECTROPHORESIS OF RNA

A sample solution with 150 µg of purified RNA was electrophoresed along with pure yeast RNA (BDH, England) acting as control on Instant Agarose gel plates (Serva, Heidelberg) for 75 minutes in veronol phosphate buffer (pH 8.2) at 300 volts and at 12 mA per gel plate. The plates were suitably stained with one per cent toluidine blue to display the nucleic acid bands. The mobilities of the RNA bands for the unknown sample were compared with those of the controls.

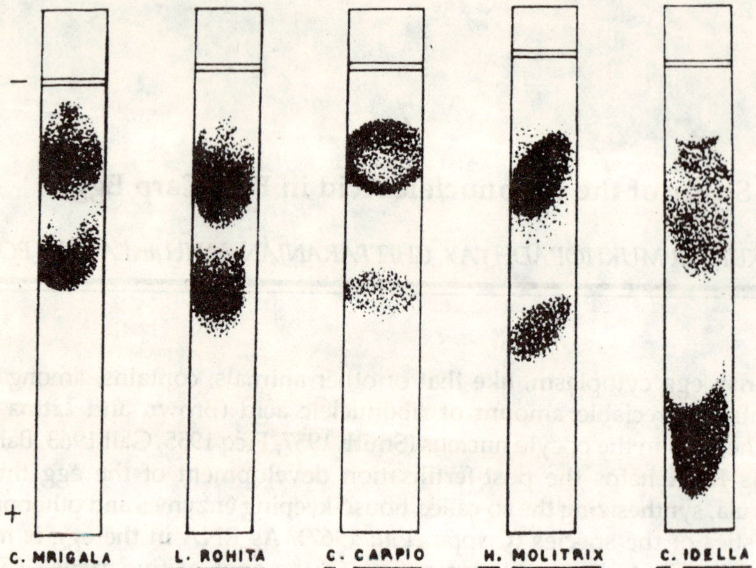

Fig. 1 : Diagrammatic Display of the Electrophoresed Agarose Gel Plate with RNA from Mature Eggs of *C. mrigala, L. rohita, C. carpio, H. molitrix* and *C. idella* (The top of the Diagram represents the origin).

Proportion of RNA Fractions in the Bands

The comparable areas of the unstained electrophoresed agarose gel plates with 150 ug of pure RNA in relation to the RNA bands were cut and the RNA in the bands was extracted through elution technique (Weime 1965, Munro and Fleck 1966) to study the relative concentrations of the RNA fractions. All the experiments were repeated five times.

RESULTS AND DISCUSSION

In all the fish eggs studied, two bands from the RNA sample are displayed. The mobilities of these bands, however, differ widely ($P<0.001$) (Table 1). The mobilities of both band I and band II are highest in *C. idella* (The Rd values 0.74 and 1.85 respectively) and the lowest in *C. mrigala* (Rd values 0.44 and 0.92 respectively). In *C. carpio* also band I has the Rd value of 0.44.

Table 1 : Mobility (Rd) Values of the Nucleic Acid Bands Separated on Agarose Gel Electrophoregrams

RNA Bands	*C. mrigala*	*L. rohita*	*C. carpio*	*H. molitrix*	*C. idella*
Ist	0.44	0.51	0.44	0.64	0.74
IInd	0.92	1.13	1.15	1.44	1.85

The concentrations of RNA fractions in the bands are not the same (Table 2). Except in *C. idella*, band I always contains a greater amount of nucleic acid fraction than band II.

Table 2: Proportion of the Nucleic Acid Fractions in Two Bands separated on Agarose Gel Electrophoregrams

RNA Bands	*C. mrigala*	*L. rohita*	*C. carpio*	*H. molitrix*	*C. idella*
Ist	5	2	11	11	1
IInd	4	1	3	4	6

In all the five species of carps only two widely separated RNA fractions could be demonstrated. This suggests a relatively homogeneous nature of RNA in the carp eggs. This was also recorded in other fish eggs (Kafiani 1970).

As RNA of a species depends entirely on its genome, the nature of this nucleic acid should differ from one species to other. This is also demonstrated in the present study. All the RNA samples from the five carps present two fractions but the relative mobility of these bands differs widely in all the species, showing a difference in the nature of the nucleic acids in the bands. Moreover, the relative concentrations of RNA in these two bands are not the same. Also, as the mobility of band I RNA fraction increases, that for band II also increases, except in *C. carpio*. At the same time, the relative concentration of RNA fraction in band I is mostly greater than that in band II (except in *C. idella*). On these two parameters, particularly on the quality of the RNA fraction, the five carps have been arranged in Table 1 and in Table 2. This difference in the RNA pattern may thus be related to a slow change in the genome of the carp group through adaptation to the environment, as suggested by Denton (1973), and helping in speciation.

REFERENCES

Balinsky, B.I. 1970. *An introduction to embryology*. (Third Edn). W.B. Saunders Company, Philadelphia.

Brown, D.D. and Littna, E. 1964. Variations in the synthesis of stable RNAs during oogenesis and development of *Xenopus laevis*. *J. Mol. Biol.*, **20** : 95-112.

Brown, D.D. and Littna, E. 1966. Synthesis and accumulation of DNA-like RNA during embryogenesis of *Xenopus laevis*. *J. Mol. Biol.*, **20** : 81-94.

Crippa, M., Davidson, E.H. and Mirsky, A.E. 1967. Persistence in early amphibian embryos of informational RNAs from the lampbrush chromosome stage of oogenesis. *Proc. Natl. Acad. Sci. U.S.*, **57** : 885-892.

Denton, T.E. 1973. *Fish chromosome methodology*. Charles Thomas Pub. U.S.

Ficq, A. 1955. Etude autoradiographique du metabolisme de oocyte de Asterius rubesn on cours de la croissance. *Arch. Biol.*, **66** : 509-524.

Gall, J.G. 1963. Chromosomes and cytodifferentiation. In : *Cytodifferentiation and macromolecular synthesis*. Ed. M. Lockke, Acad. Press, N.Y., pp 119-143.

Kafiani, C. 1970. Genome transcription in fish development. In : *Advances in morphogenesis*. Ed. M. Abercrombie and J. Brachet. Vol. 8, 209-284.

Munro, H.N. and Fleck, A. 1966. The determination of nucleic acids In : *Methods of Biochemical Analysis*. Ed. D. Glick. Interscience Publishers., N.Y., Vol. XIV, p. 113.

Sackin, M.S. 1969. Application of cross-association to an evolutionary study of cytochrome C. In. *Numerical taxonomy*. Ed. A.S. Cole. Acad. Press, N.Y., pp. 241-256.

Simpson, G.G. 1961. *Principles of animal taxonomy*. Columbia University Press, N.Y.

Smith, S. 1957. Early development and hatching. In : *The physiology of fishes*. Ed. M.E. Brown. Acad. Press. N.Y. pp. 323-359.

Tsuyuki, H. and Roberts, E. 1966. Interspecies relationships within the genus *Oncorhynchus* based on biochemical systematics. *J. Fish. Res. Canada*. **23**, 1 : 101-105.

Weime, R.J. 1965. *Agar Gel Electrophoresis*. Elsevier, Amsterdam.

Zschile, F.P. Jr. and Murray, H.C. 1963. A chromatographic spectrophotometric method for determining nucleotide composition of RNA in plant. *Plant Phy*, **38** (2) : 133-140.

Rec. Adv. Fish Ecol. Limn. Eco-conserv.
IV : 37-54

8

Limnological Studies on a Dimictic Himalayan Lake

A. WANGANEO, R. WANGNEO and D.P. ZUTSHI

INTRODUCTION

Present study was conducted during the year 1978 on Nilng Lake, (2180 m a.s.l) where from true dimictic lakes start. This water body is situated in the footholds of the Pir-Panjal range towards the south of Srinagar. The lake has an area of 6.3 ha. and is only 6.5 m deep. The lake remains frozen for 2-3 months during winter. The mean winter temperature remains well below the freezing point. This lake has no regular inflow. It receives maximum water supply through precipitation besides a few springs in the catchment area. The outflow operative during high water level is regulated by a sluice gate towards the northern end. Towards the eastern side of the lake there is a thick coniferous forest, while its western side has been opened up, where agriculture practices are in operation. From this area lot of eroded material finds its way into the lake, thus threatening the very existence of this important tourist resort. The lake supports rich floating macrophytic vegetation (*Polygonum amphibium, Nymphoides peltatum* and *Trapa natans*) towards its littoral regions.

MATERIALS AND METHODS

Monthly water sampling was conducted from a pre-fixed sampling station. The water samples were taken from different depths using a Ruttner sampler. A standard Secchi disc was used for determining visibility. Direct reading meters were used for measuring pH and conductivity while water samples were analysed following the procedure as given in APHA (1985). Plankton samples were preserved in 5 per cent formaline. Enumeration of plankton sample was done as given in Wanganeo and Wanganeo (1991).

RESULTS AND DISCUSSION

(a) Physico-chemical Characteristics

Water temperature at the surface ranged from 3 to 25°C. Minimum temperature was recorded in the month of December, giving rise to inverse stratification. Thereafter, a brief isothermal period was observed in the month of January. The whole column mixed at 4°C. This condition prevailed up to March, though slight increase in column temperature was noted. The increase was by 4°C. Thus iso-thermal period continued from January (column temp. 4°C) to March (column temp. 8°C). Further, with advancement in seasons towards summer a corresponding change in water temperature was recorded, giving rise to different temperature profiles. No stable thermocline was discernable. The maximum difference between surface and bottom temperatures was approximately 6°C (in the month of July) which gradually reduced towards the fall, when again an iso-thermal period was observed in the month of November (column temp. 6.5°C). Based on temperature regimes the lake can be categorised among dimictic lakes. Thus an elevation of > 2000 m a.s.l marks the actual zone of dimictic lakes (Fig. 1).

The Secchi transparency varied from 1.5 to 4 meters. Maximum Secchi value was recorded during the winter months (January) and minimum was recorded in October and April. The low transparency values can be attributed to the incoming silt from the catchment area on account of precipitation during Autumn and Spring season (Fig. 1).

Dissolved Oxygen in the column ranged from 0.48 mg/l to 13.2 mg/l. Throughout the year a high oxygen concentration in the upper waters in comparison to the bottom waters was observed. The maximum difference of 7 mg/l between the surface and the bottom waters was observed in the month of March. Near anoxic condition of hypolimnion has been related to the decomposing organic material which finds its way from the nearby forested area, besides the macrophytes which are abundantly present in this lake.

Conductivity in the column ranged from 190 µS to 377 µS at 25°C. In October the minimum conductivity value was observed throughout the column and the maximum value was observed in the bottom waters in the month of July. Bottom waters generally recorded high conductivity values in comparison to middle and surface waters.

Except summer, pH recorded no significant variation in the column. During summer months a maximum value of 8.6 units was recorded in the surface waters, while bottom water recorded a value of 6.9 units only. The lake water remained near neutral to alkaline throughout the study period (Fig. 2). Summer increase in pH value has been related to the photosynthetic activity (Wanganeo 1984).

In the column the carbonate ions were nearly absent throughout the year (except during summer months) and, as such, total alkalinity was mainly due to the bicarbonates. The bicarbonates throughout the column ranged from 21 mg/l to 187 mg/l. Maximum bicarbonate-ion-concentration was recorded in the month of July in the bottom waters. Generally it showed an increase with increase in the depth (Fig. 2). The bicarbonate ions dominated over the chloride ions. Chloride content ranged from 6 mg/l to 37 mg/l. Both the minimum and the maximum values were recorded from middle waters. In November the whole column recorded a uniform value of 7 mg/l.

Mg hardness was lower than Ca hardness. Mg hardness ranged from 0.72 mg/1 to 24.5 mg/1, whereas Ca hardness ranged from 20.4 mg/1 to 52.4 mg/1. Minimum values of Ca and Mg hardness throughout the column were recorded

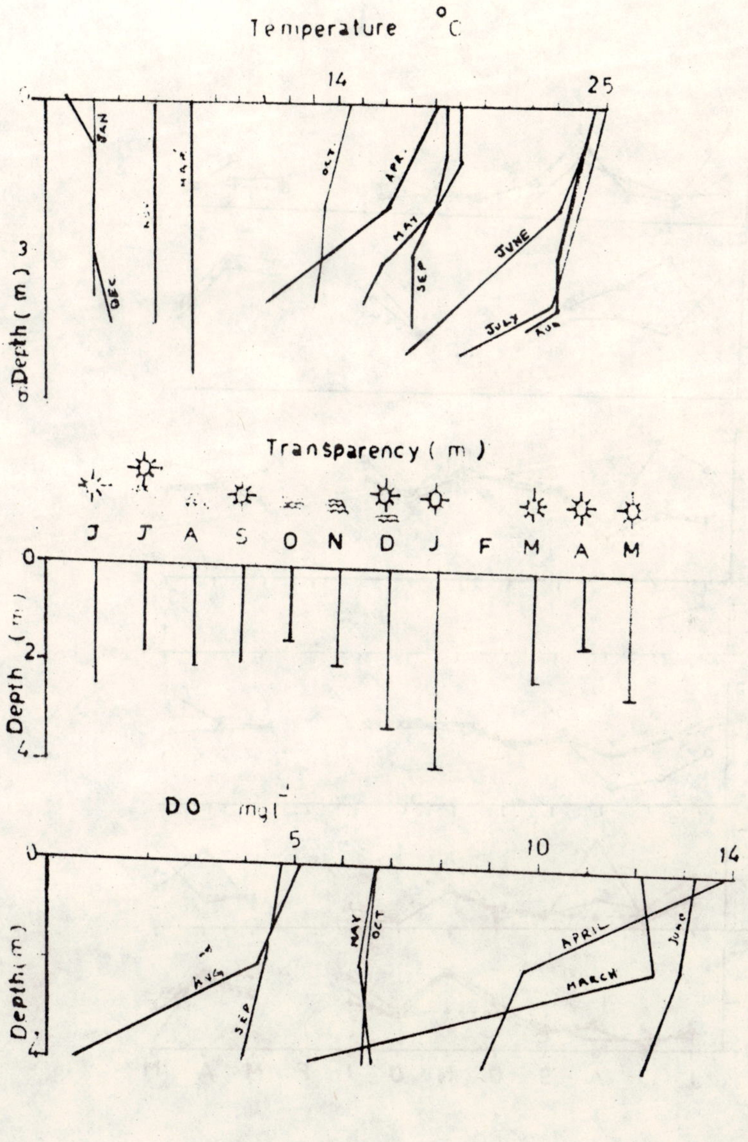

Fig. 1 : Monthly column variation in water temperature, Secchi transparency and dissolved oxygen in Nilnag lake.

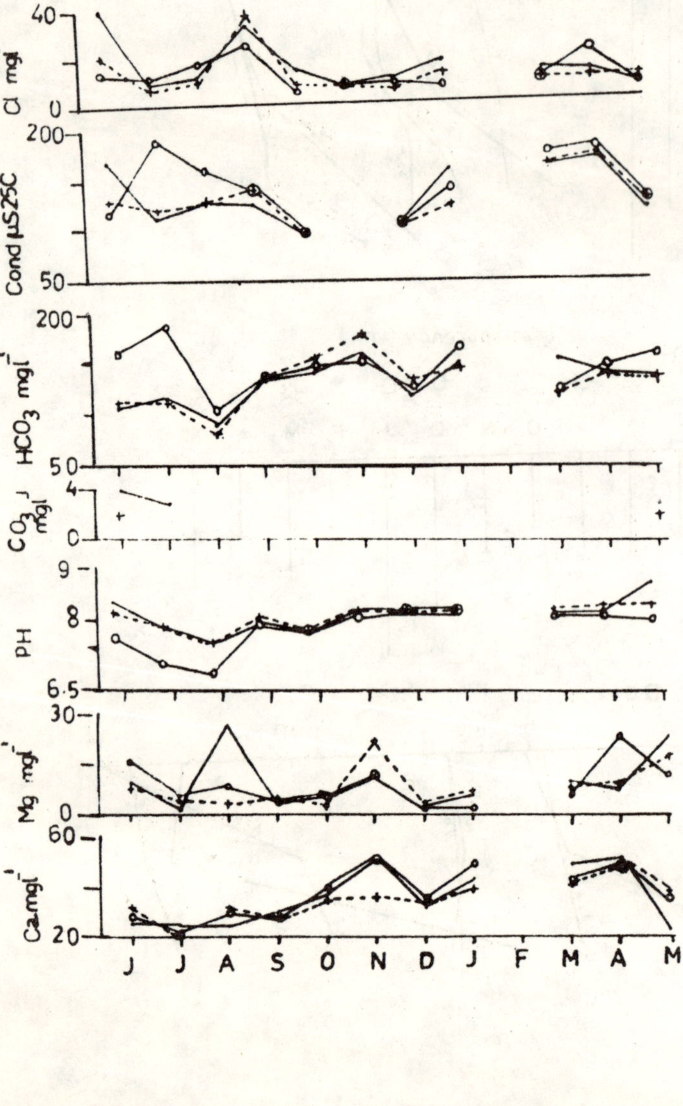

Fig. 2 : Column variation in various chemical parameters on monthly basis.
—— = Surface waters; x----x =2 meter depth; o----o = 4 meter depth.

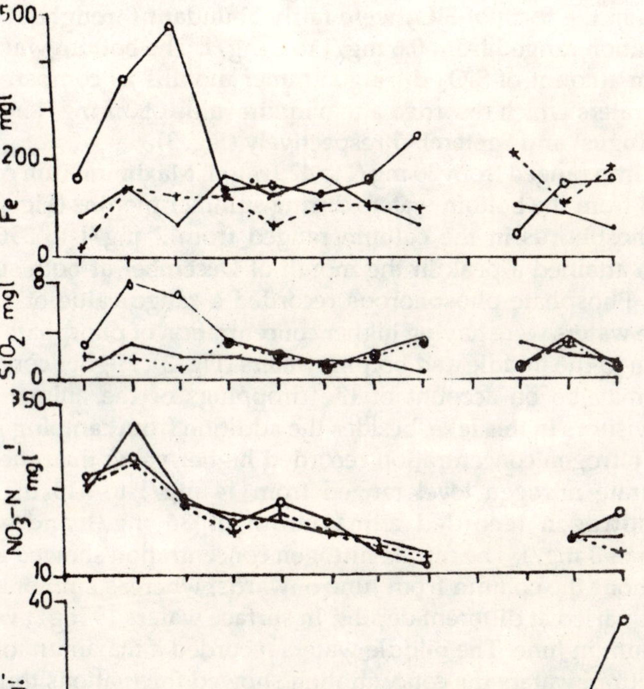

Fig. 3 : Monthly variation in important chemical parameters in a column.
—— = Surface; x----x = 2 meter depth; o——o = 4 meter depth.

in the month of July, whereas the maximum Mg hardness was recorded in May, and Ca hardness was observed in April (Fig. 2).

Silicates in the form of SiO_2 were fairly abundant throughout the column. The concentration ranged from 0.6 mg/l to 8 mg/l. The bottom waters recorded the maximum amount of SiO_2 during summer months as compared to middle and surface waters which recorded a maximum value of 3.2 mg/l in January and 2.2 mg/l in August and September respectively (Fig. 3).

Soluble iron ranged from 20 mg/l to 470 mg/l. Maximum iron concentration was recorded from the bottom waters during summer months (Fig. 3).

Total phosphorus in the column ranged from 7 mg/l to 396 mg/l. The concentration attained a peak in the month of December at both surface and 2 meter depth. Phosphate phosphorous recorded a range value of 1 mg/l to 44 mg/l. Surface waters were having higher concentration of phosphate phosphorus in comparison to the middle and bottom waters (Fig. 3). Higher concentration of phosphorus may be on account of the droppings of the animal population, carrying the visitors in this lake, besides the addition from camping grounds.

Nitrate nitrogen concentration recorded higher value than the ammonical nitrogen. Nitrate nitrogen level ranged from 14 mg/l to 313 mg/l, whereas ammonical nitrogen recorded a maximum of 36 mg/l and a minimum concentration of 3 mg/l. The nitrate nitrogen concentration showed a decreasing trend throughout the column from June onwards, whereas ammonical nitrogen concentration varied at different depths. In surface waters 19 mg/l was recorded as the maximum in June. The middle waters recorded a maximum of 10 mg/l in August. In bottom waters the concentration showed fluctuations throughout the year, attaining its maximum during the month of May (Fig. 3).

(b) Biological Characteristics

(i) Phytoplankton

During the present study a total of 86 species have been recorded throughout the year, out of which 39 species belonged to the class chlorophyceae, 34 species to the class bacillariophyceae, 6 species were recorded from the class cyanophyceae, 3 species each from dinophyceae and euglenophyceae and only one species was recorded from the class chrysophyceae. A maximum of 68 species were recorded from surface, while a minimum of 41 species were present at 5 meter depth (Table 1). Table 2 records the month of maximum occurrence of

Table 1 : Contribution of Various Algal Species in the Column on Yearly Basis

	No. of Species at Specific Depths						*Total No. of Species*
Class	*Surface*	*1m*	*2m*	*3m*	*4m*	*5m*	*Throughout the Year*
Bacillariophyceae	32	28	28	28	30	22	34
Chlorophyceae	24	15	23	24	19	10	39
Cyanophyceae	6	5	6	6	5	3	6
Dinophyceae	2	1	1	1	1	2	3
Chrysophyceae	1	1	1	1	1	1	1
Euglenophyceae	3	3	3	3	3	3	3

Table 2 : Contribution of Various Algal Species in the Column on Yearly Basis

	0 m		*1m*		*2m*		*3m*		*4m*		*5m*	
	Month of Maximum Occurrence	*Abundance*	*Month of Maximum Occurrence*	*Abundance*	*Month of Maximum Occurrence*	*Abundance*	*Month of Maximum Occurrence*	*Abundance*	*Month of Maximum Occurrence*	*Abundance*	*Month of Maximum Occurrence*	*Abundance*
BACILLARIOPHYCEAE												
Achnanthes sp.	June	F	July	S	June	F	June	F	July	F	June 1	R
Amphora sp.	May	F	July	S	July	R	August	F	July	F	June 1	F
Amphora bitumida	Aug.	R	Jan.	R	Sep.	R	—	—	—	—	June 1	R
Cocconeis placentula	April	F	July	F	June/Nov	R	Jan	R	July	F	June 1	R
Cyclotella sp.	April	F	July	F	June/Nov	R	Jan	S	Aug.	F	June	R
Cymatopleura solea	Dec	R	July	F	—	—	Aug	R	July	R	June 1	R
Cymbella	April	F	July	S	April	F	Jan.	F	July	S	June 1	S
Diatoma elongatum	June 1	F	July	F	June	F	June	F	July	F	June 1	F
D. hiemale	—	—	Sep.	R	—	—	—	—	—	—	—	—
Epithemia turgida	Oct	R	July	F	Aug	R	Aug	R	Nov	F	June	R
Eunotia pectinalis	April	F	Dec.	F	July	R	Jan	F	July	F	June 1	F
F. arcus	March	F	July	F	July/Apr	R	June	F	Mar	S	—	—
F. capucina	Jan	F	—	—	Nov	F	Jan	S	June	F	—	—
Fragillaria crotonensis	March	D	July	D	July	F	July	S	July	S	June 1	F
Gomphoneis herculeanum	Aug	R	July	F	—	—	June	R	July	R	—	—
G. accuminatum	June 1	R	July	R	Dec.	R	—	—	June 1	R	June 1	R
G. geminatum	March	R	July	S	June/Nov	R	July	F	June 1	F	June 1	R
G. olivaceum	June	R	July	F	Mar	R	June	R	July	F	June 1	F
G. subtile	June 1	R	—	—	—	—	—	—	May	R	—	—
Gyrosigma sp.	Oct.	R	July	F	Nov.	R	Nov.	R	Mar	S	June 1	F
Hantzschia sp.	April	R	Dec.	R	Aug	R	May	R	Oct	R	June 1	R
Melosira sp.	June/Sep.	R	July	S	June	R	June	R	July	F	—	—
Navicula confervacea	April	F	—	—	July	R	Jan	S	—	—	—	—
N. cuspidata	April	R	Jan.	R	July	R	Nov	F	Sep	F	June 1	R
Navicula sp.	Mar./May	F	July	S	Nov	F	Jan/July	F	July	S	June 1	F
Nitzschia accicularis	July	D	Aug	S	July	D	July	D	July	S	June 1	—
N. linearis	June 1	R	June 1	R	—	—	—	—	June 1	R	—	—

contd. . . .

	0m		1m		2m		3m		4m		5m	
	Month of Maximum Occurrence	*Abun-dance*	*Month of Maximum Occurrence*	*Abun-dance*	*Month of Maximum Occurrence*	*Abun-dance*	*Month of Maximum Occurrence*	*Abun-dance*	*Month of Maximum Occurrence*	*Abun-dance*	*Month of Maximum Occurrence*	*Abun-dance*
N. sigmoides	June 1	R	—	—	Mar	R	Nov	R	Mar	F	—	—
N. corenziana	—	—	—	—	—	—	—	—	—	—	June 1	R
Pinnularia sp.	April	R	Sep	R	Sep	R	Nov	R	Nov	R	—	—
Rhizosolenia sp.	Aug.	R	—	—	Aug	R	July	R	Aug	R	—	—
Rhopalodia gibba	July	F	Jan	F	July	R	Nov	R	June/July	R	June	F
Suirella ovalis	May	R	July	R	Oct	R	June	R	July	R	—	—
Syndera ulna	June	F	July	S	June	F	Aug	F	Mar	S	June	R
CHLOROPHYCEAE												
Ankistrodesmus falcatus	—	—	—	—	June	R	Aug	R	June	R	—	—
A. spiralis	July	R	—	—	June	R	—	—	—	—	—	—
Arthrodesmus sp.	Aug.	R	—	—	—	—	—	—	—	—	—	—
Arthrospira sp.	Oct	R	—	—	Oct	R	July	R	Nov	S	—	—
Closteriopsis longissima	June	R	—	—	—	—	—	—	—	—	—	—
Closterium sp.	Mar	R	July	R	—	—	—	—	—	—	June	R
Coelastrum sp.	Aug.	R	July	R	Aug	R	June/Aug	R	July	R	—	—
Cosmarium hiamalis	June	R	—	—	—	—	Aug	R	—	—	—	—
C. reniforne	—	—	—	—	July	R	—	—	—	—	—	—
Cosmarium sp.	June	F	July	F	June	R	June	R	July	F	June	F
Crucigenia crucifera	Aug	R	Sep.	R	Aug/Sep	R	Aug	F	Aug	F	—	—
C. tetrapedia	—	—	—	—	—	—	June	R	Aug	F	June	R
Desmidium sp.	Aug	R	—	—	Nov	F	—	—	—	—	—	—
Euastrum sp.	—	—	July	R	—	—	—	—	Aug	R	—	—
Gomphosphaeria sp.	—	—	—	—	July	R	June	R	—	—	—	—
Oedogonium sp.	Jan	R	Oct	R	Sep	F	Sep	R	Aug	R	June	R
Pediastrum duplex	—	—	—	—	—	—	—	—	—	—	June	R
P. tetras	Aug	R	June	R	—	—	Aug	R	June/Sep	R	—	—
Scenedesmus arcuatus	Jul/Aug/Sep	R	July	R	Aug	F	July	F	Aug	R	—	—
S. armatus	—	—	—	—	Sep	R	Aug	R	Aug	R	—	—
S. bijugatus	Aug.	R	July	R	July	F	Aug	R	Aug	F	—	—
S. dimorphus	Aug	R	—	—	Nov	R	Aug	R	—	—	—	—
S. quadricauda	—	—	—	—	Sep	R	—	—	Aug	R	—	—

Selenastrum sp.	—	—	—	—	—	—	Aug	R	—	—	—	—
Sphaerocystis schroeteri	June	R	—	—	—	—	July	F	—	—	—	—
Sphaerozosma acbertianum	Aug	F	July	R	Aug	F	Aug	R	July	F	June	F
Spirulina sp.	—	—	July	R	—	—	Sep	R	Nov	F	June	R
Spondylosium sp.	Jan	R	—	—	—	—	—	—	—	—	—	—
Staurastrum appliculatum	—	—	—	—	June	R	—	—	—	—	—	—
S. bienecius	—	—	June	F	June	R	—	—	—	—	—	—
S. lunatum	June	R	—	—	—	—	—	—	—	—	—	—
Saturastrum sp.	July	F	July	F	July	F	July	F	July	F	July	R
Staurodesmus sp.	April	F	July	R	Mar/Apr/	R	June	R	June	R	June	R
Tetradon minium	June	F	Jun July	F	Aug.	F	June	S	July	S	June	R
T. muticum	—	—	—	—	—	—	Aug	R	—	—	—	—
T. pusillum	July	R	—	—	Aug/Sep	R	July	R	—	—	—	—
T. trilobatum	—	—	—	—	Aug	R	—	—	—	—	—	—
Tetrastrum sp.	—	—	—	—	—	—	Aug	R	Aug	R	—	—
Treubaria triapend	June	R	—	—	—	—	—	—	—	—	—	—
CYANOPHYCEAE												
Anabaena sp.	July	S	Aug.	R	July	R	Aug	R	July	F	July	R
Aphanocapsa sp.	June	R	—	—	June	R	Aug	R	—	—	—	—
Coelosphaerium sp.	Aug.	R	July	R	Aug	F	July	F	July	F	—	—
Merismopedia sp.	July	F	Sep	R	Sep	F	July	F	Sep	F	—	—
Microcystis sp.	Jan	R	July	R	April	R	May/June	R	Aug	F	June	R
Oscillatoria sp.	June	R	Nov	F	June	R	Sep	R	Nov	S	June	S
DINOPHYCEAE												
Glenodinium sp.	Aug.	R	—	—	—	—	—	—	—	—	—	—
Gymnodinium sp.	—	—	—	—	—	—	—	—	—	—	June	S
Peridinium sp.	Oct.	F	Sep/Oct	F	July	S	July	S	July	S	June	S
CHRYSOPHYCEAE												
Dinobryon sp.	June	F	July	S	July	S	June	S	June	S	June	S
EUGLENOPHYCEAE												
Euglena acus	June	F	June	R	June	F	July	F	July	S	June	S
Euglena sp.	Sep.	R	June	F	July	F	July	S	July	S	June	F
Phacus sp.	June	R	Sep	R	July	F	July	S	July	S	June	S

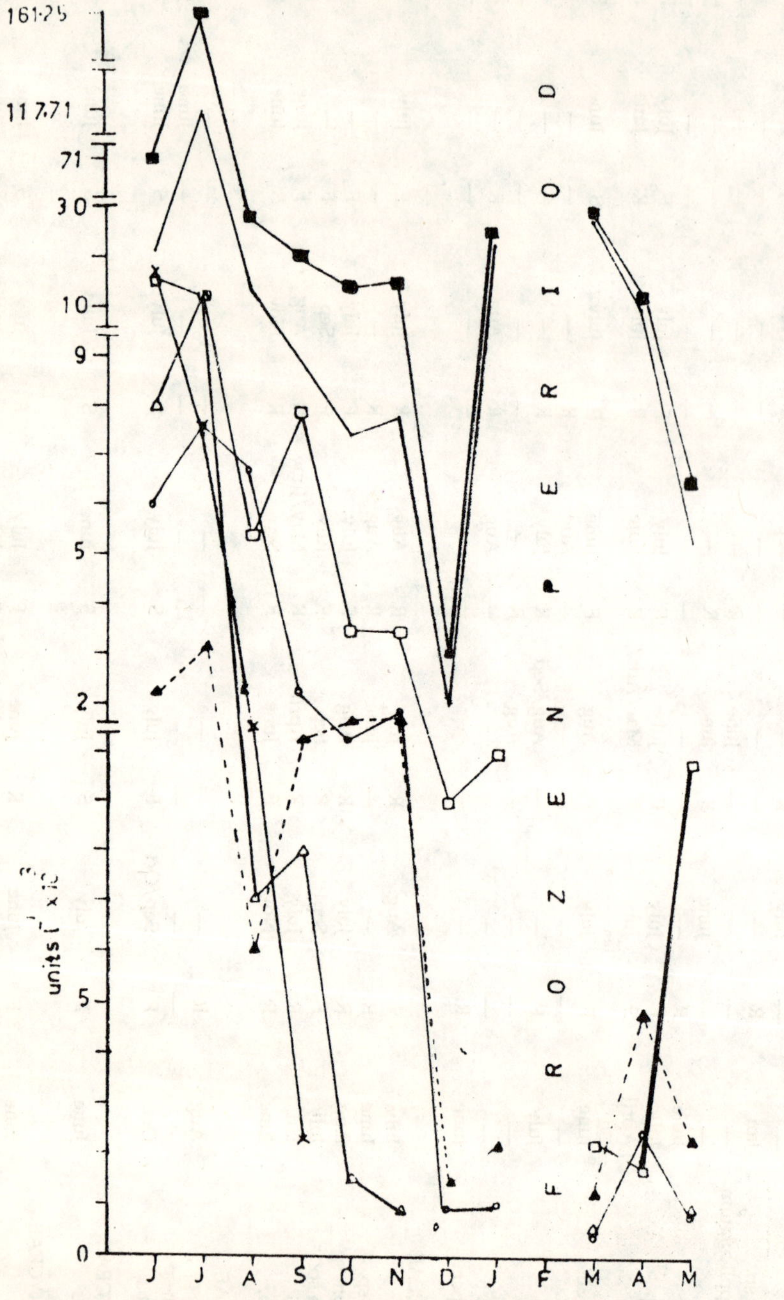

Fig. 4 : Monthly variation in various algal classes in Nilnag lake on column average basis. ——— = Bacillariophyceae; --△-- = cyanophyceae; ---x--- = Chrysophyceae; ---o--- = Chlorophyceae; —□— = Dinophyceae; --▲-- = Euglenophyceae; —■— = Total algae.

individual species. The total phytoplankton density of the column throughout the year attained two peaks, the highest in the month of July (161.25 x 10 units/l) and a smaller one in the month of March (28.16 x 10 units/l). Amongst various classes in the column, bacillariophyceae also showed two peaks, one in July (117.71 x 10^3 units/l) and the other in March (27.7 x 10^3 units). Chlorophyceae, cyanophyceae, dinophyceae, chrysophyceae and euglenophyceae each recorded only one peak, contributing 7.65 x 10 units/l (July), 3.16 x 10^3 units/l (July), 15.57 x 10^3 units/l (June), 17.25 x 10^3 units/l (June) and 18.24 x 10^3 units/l (July) respectively (Fig. 4).

The average percentage contribution of various phytoplankton classes at various depths show that Bacillariophyceae was the most dominant group from surface to 4 meter depth, contributing about 88 per cent, 79 per cent, 63 per cent, 64 per cent and 49 per cent respectively. At 5 meter depth dinophyceae was the dominant class, contributing 34 per cent of the total phytoplankton population (Table 3).

Table 3: Average Percentage Contribution of Various Algal Classes in the Column (Units/l)

Depth (m)	*Bacillari-phyceae*		*Chlorophy-ceae*		*Cyanophy-ceae*		*Dinophy-ceae*		*Chryso-phyceae*		*Eugleno-phyceae*	
	u/l	%	*u/l*	%	*u/l*	%	*u/l*	%	*u/l*	%	*u/l*	%
0	5417	87	268	4	194	3	161	3	153	2	57	1
1	3442	79	270	6	83	2	265	6	282	6	24	1
2	2693	63	308	7	83	2	477	11	559	13	179	4
3	4207	64	483	8	82	1	1012	15	361	5	429	7
4	3926	49	786	10	450	6	1372	17	668	8	758	10
5	867	32	87	3	104	4	922	34	253	10	446	17

Bacillariophyceae attained maximum population in the month of July from surface to 4 meter depth. The density at various depths ranged from 18.08x10^3 units/l to 31.95x10^3 units/l, contributing more than 53 per cent at all the depths. However, at 5 meter depth a maximum of 1.57 x 10^3 units/l (33 per cent) was recorded in the month of June (Fig. 5). Wanganeo and Wanganeo (1991) recorded the dominance of class bacillariophyceae in the lakes above 3000 m a.s.l, contributing approximately 80 per cent towards the phytoplankton, while in lakes below 2000 m a.s.l., a contribution of only 40 per cent was recorded by the authors.

Chlorophyceae attained a maximum population in the month of June at surface, 3 and 5 meters in July at 1 and 4 meters and in August at 2 meter depths. The number ranged from 0.88 x 10^3 units/l to 2.43 x 10^3 units/l, contributing more than 2 per cent at all the depths (Fig. 5)

Cyanophyceae attained a maximum population in July at surface and 3 meter; in November at 1 meter; in September at 2 meter, in October at 4 meter and in June at 5 meter depths. The number ranged from 0.24 x 10^3 units/l to 1.8 x 10^3 units/l at various depths, contributing more than 2 per cent at all the depths. Dinophyceae attained a maximum population in July at 2, 3 and 4 meters, in October at surface, in September at 1 meter and in June at 5 meter depths. The number at various depths ranged from 0.82 x 10^3 units/l to 11.07 x 10^3 units/l, contributing more than 5 per cent at all the depths (Fig. 5). Chrysophyceae

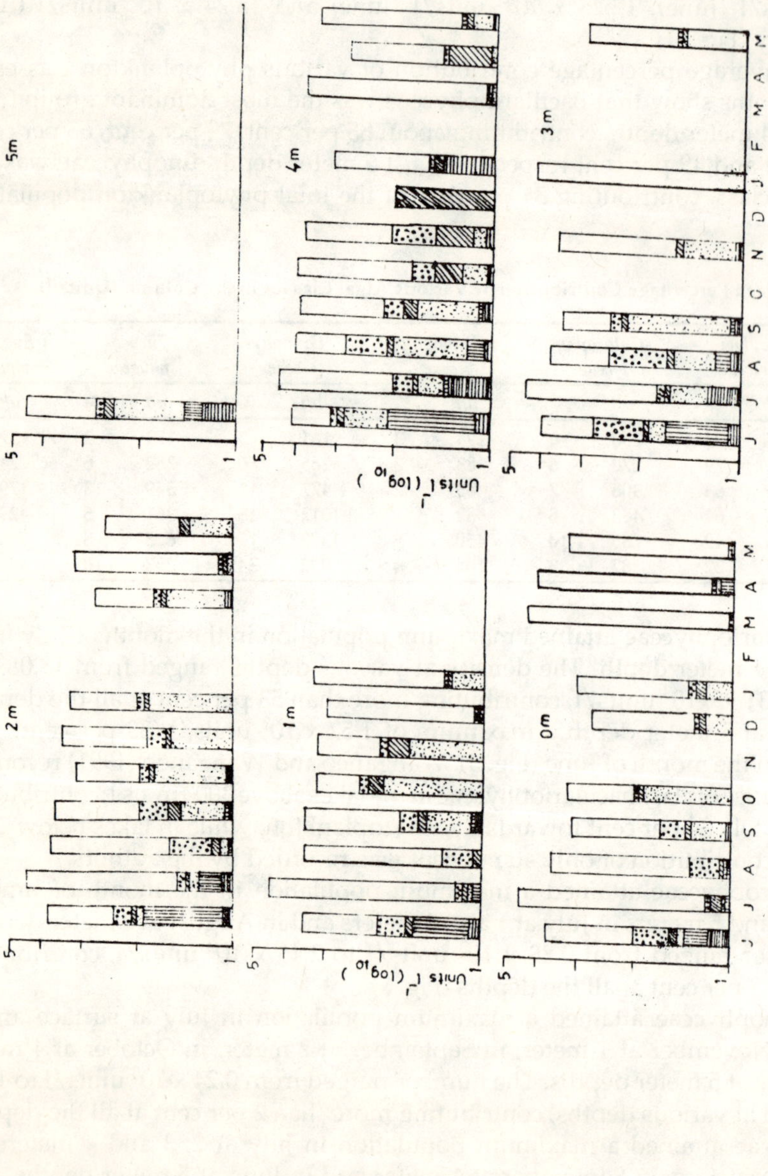

Fig. 5 : Monthly per cent contribution of various algal classes at different depths in Nilnag lake. = Bacillariophyceae; = Chlorophyceae; = Cyanophyceae; = Dinophyceae; = Chrysophyceae; = Euglenophyceae

attained a maximum population in June at surface, 3, 4 and 5 meters, in July at 1 and 2 meter depths. The number ranged from 0.72×10^3 units/l to 6.39×10^3 units/l, contributing more than 6 per cent at all the depths (Fig. 5). Euglenophyceae attained a maximum population in June at surface 1 and 5 meter depths, in July at 3 and 4 meters. The number ranged from 0.20×10^3 units/l to 6.72×10^3 units/l, contributing more than 5 per cent at all the depths. The sequence of dominance among phytoplankton classes in the lake was:

Bacillariophyceae>Dinophyceae>Chrysophyceae>Chlorophyceae> Euglenophyceae>Cyanophyceae.

But in surface water bacillariophyceae was followed by chlorophyceae; this can be attributed to the fact that members of dinophyceae are motile and, as such, avoid high intensity of light (Wanganeo and Wanganeo 1991). Cyanophyceae, which normally followed bacillariophyceae in the valley lakes of Kashmir Himalayas situated at an altitude of approx. 1500 m a.s.l (Wanganeo and Wanganeo 1991), was the least represented class here.

The population density showed a gradual increase from winter through spring to summer, where it attained a uni-modal peak both in case of Phyto as well as the Zooplankton, though in case of Phytoplankton a spring impulse was also recorded. The population showed a decline during the winter season.

All algal classes attained their maximum during summer period only. This clearly indicates that the algal populations present in the lake under study are temperature dependent. Wanganeo and Wanganeo (1991) reported that that lakes situated at an average altitude of 1500 m a.s.l depicted varied types of algal patterns, because of their different morphological and chemical characteristics. Kant and Kachroo (1977) reported maximum phytoplankton density in Autumn in Dal and Nagin lakes of Kashmir. Khan (1978) obtained unimodal behaviour of phytoplankton in Naranbagh lake, Kashmir. Raina (1981) observed two peaks, one in winter and another in summer in Bod-Sar, Kashmir. Only two species were found to be dominant which belonged to the class bacillariophyceae. *Nitzschia accicularis* was abundant throughout the year and at almost all the depths. It attained its maximum in the month of July at surface and 2 meter depths. At surface its contribution to the class is 94 per cent and the total phytoplankton 83 per cent. At 2 meter depth its contribution to the class is 92 per cent and to the total phytoplankton it is 67 per cent (Table 2). *Fragillaria crotonensis* was also present almost throughout the year. It was found to be abundant at surface in March and at 1 meter in the month of July. At surface, its contribution to the class was 90 per cent and it contributed 67 per cent to the total phytoplankton population. At one meter depth its contribution was 42 per cent to the class, whereas it contributed 37 per cent to the total phytoplankton population (Table 2). An increase in *Fragillaria crotonensis* suggests a response to nutrient levels and was associated with the eutrophication of lake Washington (Stockner and Benson 1967). William (1969) reported that *Fragillaria* sp. shows up if sewage enters the lake. Jarnefelt (1952), Rawson (1956) and Vollenweider (1970) also considered *F. crotonensis* to indicate eutrophy.

(ii) Zooplankton

Only 14 species were recorded from the group throughout the year, out of which 9 species were from rotifera, 2 belonged to cladocera and 3 species belonged to copepoda.

Table 4 : Contribution of Various Zooplankton Species in the Column on Yearly Basis

Class	*No. of Species at Specific Depths*						Total No. of Species
	Surface	*1 m*	*2 m*	*3 m*	*4 m*	*5 m*	*Throughout the Year*
Rotifera	5	3	4	4	7	4	9
Cladocera	-	1	1	1	2	1	2
Copepoda	1	1	2	3	2	3	3

A maximum of 11 species was recorded from 4 meter depth and a minimum of 6 species from surface waters (Table 4). Monthly occurrence of various zooplankton species is given in Table 5. The total zooplankton density attained a single peak of 12.19 x 10 ind./l in the month of June. Vasisht and Sharma (1975), Gophen (1972), Wanganeo (1980) and Raina (1981) reported bimodal pattern of zooplankton in freshwaters, while Vass *et. al* (1989) reported unimodal behaviour in high mountain lakes. The class rotifera also attained only one peak of 11.79 x 10 ind./l during June. The classes cladocera and copepoda did not show any peak (Fig. 6). Vass *et. al* (1989), Wanganeo (1980) and Raina (1981) also reported predominance of rotifers in various Kashmir Himalayan lakes. The average percentage contribution of various classes at various depths showed that class rotifera was the most dominant group, contributing 94 per cent, 83 per cent, 82 per cent, 92 per cent, 90 per cent, and 97 per cent respectively from surface to 5 meter depths (Table 6).

Rotifera attained a maximum population in the month of July at surface, 2 and 3 meters and in June at 1, 4 and 5 meter depths. The density at various depths ranged from 0.20×10^3 ind./l to 6.18×10^3 ind./l, contributing more than 80 per cent at all the depths (Fig. 7).

Cladocera was found to be absent at the surface, whereas it attained its maximum in September at 1 meter, in June at 2 and 5 meters, in January at 3 meter; at 4 meter depth the same number was recorded in June and August which was the maximum attained at this depth. The number at various depths ranged from 0.02×10^3 ind./l to 0.06×10^3 ind./l, contributing more than 1 per cent at all depths. The general scarcity of cladocerans in lakes has been related to a number of factors, viz., shortage of suitably sized food particles and the predation by fish (Burgis 1973, Wanganeo 1980 and Raina 1981).

Copepoda attained a maximum population in May at surface and 4 meters, in July at 1, 2 and 3 meters and in June at 5 meter depths. The number at various depths ranged from 0.02 x 10 ind./l to 0.39 x 10 ind./l, contributing more than 2 per cent at all depths. The dominance of rotifers over other zooplankters depicts,

Table 5: Monthly Occurrence of Various Zooplankton Species

	Month of Maximum Occurrence	Abundance	Month of Maximum Occurrence	Abundance	Month of Maximum Occurrence	Abundance	Month of Maximum Occurrence	Abundance	Month of Maximum Occurrence	Abundance	Month of Maximum Occurrence	Abundance
	Surface		*1m*		*2m*		*3m*		*4m*		*5m*	
ROTIFERA												
Anuraepsis fissa	July	R	Aug	R	—	—	—	—	Aug	F	—	—
Brachionus angularis	Aug	R	—	—	—	—	—	—	—	—	—	—
B. caudata	—	—	—	—	—	—	—	—	Aug	F	—	—
B. quaadridentatus	—	—	Oct/Nov	R	—	—	—	—	—	—	June	R
Colurella bicuspicata	July	R	—	—	—	—	—	—	Aug.	R	—	—
Filinia longiseta	—	—	—	—	Aug	R	Aug	R	July	F	—	—
Keratella cochlearis	July	F	June	F	July	F	July	S	June	S	June	S
K. quadrata	—	—	—	—	April	F	May	R	Mar	F	June	R
Polyarthra sp.	May	R	—	—	June/Mar	R	June	R	June	F	June	F
CRUSTACEA												
Alona sp.	—	—	—	—	—	—	—	—	April	R	—	—
Ceriodaphnia sp.	—	—	Sep	R	June	R	Jan	R	June/Aug	R	June	R
COPEPODA												
Cylops sp.	—	—	—	—	Aug	R	Aug	R	May	F	June	R
Diaptomus sp.	May	R	July	R	—	—	July	R	—	—	June	R
Nauplius	—	—	—	—	July	R	July	R	May	R	June	R

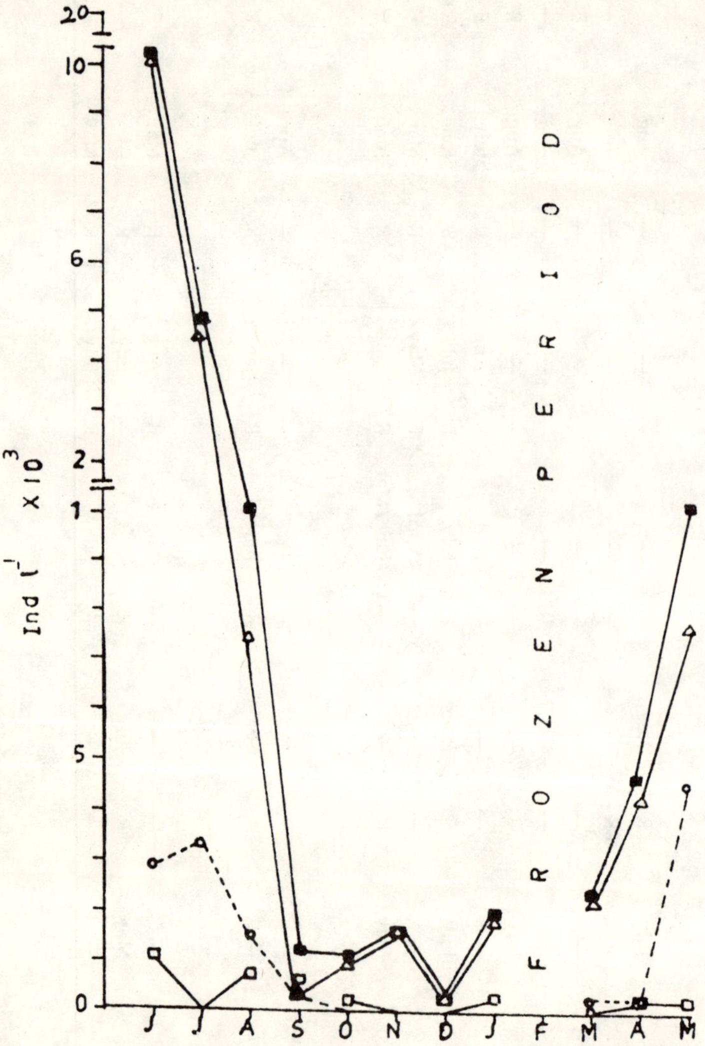

Fig. 6 : Monthly variation in various zooplankton classes in Nilnag lake on column average basis.
--■-- = Total zooplankton; --△-- = Rotifera; o----o = Cladocera; □----□ Copepoda.

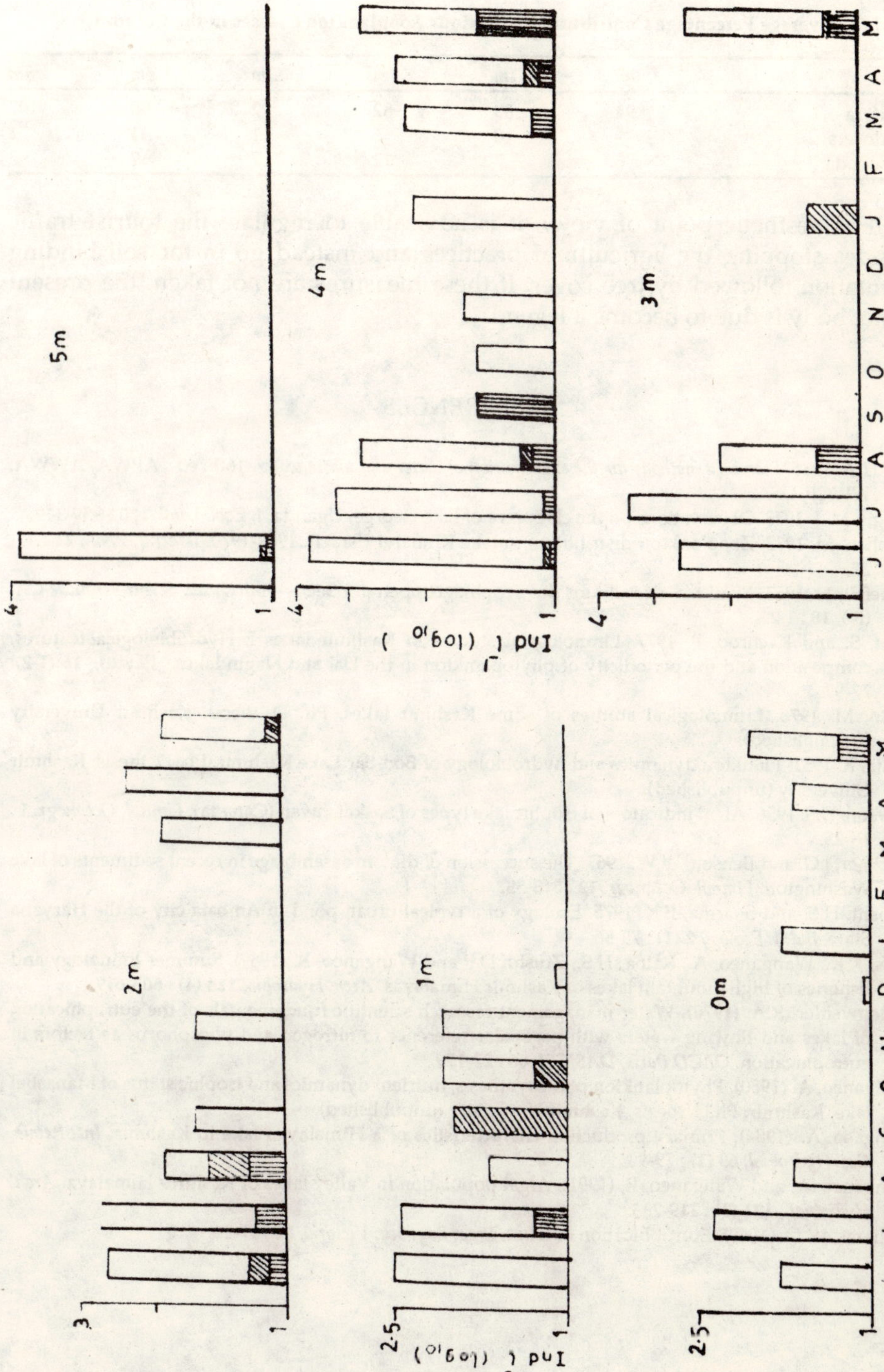

Fig. 7 : Monthly per cent contribution of various zooplankton classes at different depths in Nilnag lake. □ = Rotifera; ▨ = Cladocera; ▥ = Copepoda.

Table 6 : Average Percentage Contribution of Various Zooplankton Classes in the Column

Class	*0m*	*1m*	*2m*	*3m*	*4m*	*5m*
Rotifera	94	83	82	92	90	97
Cladocera	-	10	6	1	1	1
Copepoda	6	7	12	7	9	2

well as aesthetic point of view), it is advisable to regulate the tourist traffic besides stopping the agricultural practices and instead go in for soil binding plantation followed by tree cover. If these measures are not taken, the present water body is due to become a legend.

REFERENCES

A.P.H.A. 1985. *Standard methods for the examination of water and waste water*. 16th. ed., APWA, AWWA, WWCF, USA.

Burgis, M. J. 1973. Observations on the cladocera of lake George, Uganda. *J. Zool. Lond.* **170** : 339-349.

Gophen, M. 1972. Zooplankton distribution in lake Kinnaret (Israel). 1969-1970. *Israel. J. Zool.*, **21** : 17-27.

Jarnefelt, H. 1952. Plankton als indiktor des Trophiegruppen der Seen—*Ann. Acad. Scient Fennicae* CA, (iv), **18** : 1-29.

Kant, S. and Kachroo, P. 1977. Limnological studies in Kashmir lakes I. Hydrobiological features, composition and the periodicity of phytoplankton in the Dal and Nagin lakes. *Phykos,* **16** (1-2): 77-97.

Khan, M. 1978. Limnological studies of some Kashmir lakes. Ph. D. thesis, Kashmir University (Unpublished).

Raina, R. 1981. Plankton dynamics and hydrobiology of Bod-Sar Lake Kashmir. Ph. D. thesis, Kashmir University (unpublished).

Rawson, D.S. 1956. Algal indicators of trophic lake types of Sasketchwan (Canada). *Limnol. Oceanogr.* **1** : 18-25.

Stockner, J.G. and Benson, W.W. 1967. The succession of diatom assemblage in recent sediments of lake Washington. *Limnol. Oceanogr.* **12** : 513-532.

Vasisht, H.S. and Sharma, B.K. 1975. Ecology of a typical urban pond in Ambala city of the Haryana State. *Ind. J. Ecology* **2** (1) : 79-86.

Vass, K.K., Wanganeo, A., Raina, H.S., Zutshi, D.P and Wanganeo, R. (1989). Summer limnology and fisheries of high mountain lakes of Kashmir Himalayas. *Arch. Hydrobiol* **114** (4) : 603-619.

Vollenweider, R.A. (1970). Water management research scientific fundamentals of the eutrophication of lakes and flowing waters with particular reference to nitrogen and phosphorus as factors in eutrophication. *OECD Paris, DAS/CSI,* **68** : 27-159.

Wanganeo, A. (1980). Phytoplankton photosynthesis, nutrient dynamics and trophic status of Manasbal lake, Kashmir. Ph.D. thesis, Kashmir University (unpublished).

Wangaeo, A. (1984). Primary production characteristics of a Himalayan lake in Kashmir. *Int. Revue. Ges. Hydrobiol,* **69** (1) : 79-90.

Wanganeo, A. and Wanganeo, R. (1991). Algal population in Valley lakes of Kashmir Himalaya. *Arch. Hydrobiol.,* **121** (2) : 219-233.

Williams, W.D. (1969). Eutrophication of lakes, *Proc. Roy. Soc. Victoria,* **83** : 17-26.

Rec. Adv. Fish Ecol. Limn. Eco-conserv.
IV : 55-58

9

Rare Wildlife of Uttarkhand (India) and its Conservation

S.M. DAS

The Himalayan territory of Uttarkhand is comprised of the two divisions of Kumaun and Garhwal. The present contribution gives a review of the survey of vanishing wildlife species threatened with extinction and for which urgent conservation measures must be taken, as otherwise we will lose them for ever. The contribution also gives the latest scientific nomenclature and vernacular names along with the altitudinal ranges of various species of wild animals of the region.

Although the *Red Data Book* lists about 600 species of animals threatened with extinction all over the world, and protective measures have been taken for many of them, very few steps have been taken to conserve the unique Uttarkhand wildlife, which is on the point of vanishing altogether. This comprises of about 30 species of mammals which will disappear in the next 50 years or so, if no protection measures are adopted at the earliest. The speed of extinction is also so fast today that over 40 species have vanished during the past 50 years.

The new Wildlife Protection Act, 1972 has totally forbidden killing of the threatened species. The Indian Board of Wildlife (founded in 1952) and the Wildlife Warden with authorized forest officers are supposed to look after the enforcement of the 1972 Act to prevent trapping, shooting, poaching, serving their meat in hotels, as also prevention of export of skins, hides, furs, hair, horns and other products of the protected wildlife. But in a vast country mountainous region such as Uttarkhand the enforcement of the Act has been well nigh impossible.

The main reason for this inaction lies in the inaccessibility of the steep and high altitudes which these rare animals inhabit, usually from 6000 ft. to 18000 ft. Surprisingly in no other country of the world is life and living organisms held so sacred as in India. One may observe village people feeding ants, squirrels, birds and the like. But it is amazing that so much wildlife is still being destroyed for pleasure of the hunt, for food or for monetary gains. One finds in the Ashoka pillar edict the names of some beasts, birds and fish that are to be afforded protection. Even such ancient works as Kalidasa's *Abhayaranya* names many wildlife species which are to be given protection.

How is it that Indians have lost this heritage of respect and the faculty for protection of wildlife? They have forgotten their responsibilities for wildlife and preservation of nature, which have been replaced by greed, gain and personal aggrandisement. They have lost the faculty to appreciate that any species of wildlife is a unique creation of nature that cannot be manufactured by man. Once extinct, it is impossible to reconstruct even an ant or a tiny mouse.

Besides, wild animals not only form an essential or integral part of ecology of any hill, forest or grassland, but they are also essential for a balance of nature, without which no life including that of man can thrive.

Among the wildlife of Uttarkhand we have animals such as Thar, Goral, Chowsingha, Antelope, Chinkara, Gazella, Clouded Leopard, Caracal, Himalayan marmot, Himalayan Giant Flying Squirrel, Himalayan vole and the Himalayan mouse-hare, which are found nowhere else in the world today. Thus we have the unique task of preserving these vanishing species for the whole world to see, admire and study. Several crores of rupees are being spent for conservation of the Lion, the Tiger, the Neel-gai, the Bison, the Spotted Deer (Cheetal) and the Golden Langur. Few attempts have been made to preserve the unique forms of rare Uttarkhand wildlife species listed below :

	Common & Vernacular Name	*Old Name*	*Current Valid Name*
1.	Brown Red Bear (*Lal Bhalu*)	*Ursus isabellinus*	*Ursus arctos isabellinus*
2.	Himalayan Black Bear (*Reechh*)	*U. tibetanus*	*Selenarctos thibetanus*
3.	Himalayan Ermine (*Chitrola*)	*Mustela erminea*	*Mustela sibirica erminea*
4.	Hill Otter (*Udbilao*)	*Lutra vulgaris*	*Lutra lutra monticola*
5.	Clawless Otter	*Aonyx leptonyx*	*Aonyx cinerea concolor*
6.	Snow Leopard (*Bhural*)	*Felis uncia*	*Panthera uncia uncia*
7.	Jungle Cat (*Khattas*)	*Felis caracal*	*Felis caracal kutas*
8.	Caracal/Red Lynx (*Syahghooh*)	*F. caracal*	*F. caracal schmitzi*
9.	Clouded leopard (*Am cheeta*)	*Neofelis nebulosa*	*Neofelis neofelis macroscelloides*
10	Spotted Tiger Civet (*Zikchum*)	*Prinodon pardicolor*	*Prionodon pardicolor pardicolor*
11.	Himalayan Palm Civet (*Khattas*)	*Paguma laniger*	*Paguma larvata grayi*
12.	Indian Wolf (*Bheria*)	*Canis pallipes*	*Canis lupus pallipes*
13.	Red Hill Fox (*Lal Lomri*)	*Vulpes montana*	*Vulpes vulpes montana*
14.	Red Giant Flying Squirrel (*Lal Gilhari*)	*Pteromys petaurista*	*Petaurista petaurista*
15.	Kumaun Flying Squirrel	*Hylopetes albiniger*	*Petaurista magnificus*
16.	Himalayan Marmot (*Pahari Moos*)	*Marmota himalayanus*	*Marmota bobak himalayana*
17.	Himalayan Vole (*Pahari Chuha*)	*Arvicula royelei*	*Alticola royeli*
18.	Himalayan Porcupine (*Pahari Sahi*)	*Hystrix leucora*	*Hystrix hodgsoni hodgsoni*
19.	Himalayan Mouse Hare (*Chhota pahari Khargosh*)	*Lagomys royelei*	*Ochotona royelei royelei*
20.	Hispid Hare (*Bara Pahari Khargosh*)	*Lepus hispidus*	*Caprolagus hispidus*
21.	Musk Deer (*Kasturi*)	*Moschus moschiferus*	*Moschus moschiferus moschiferus*
22.	Four-horned Antelope (*Chausingha*)	*Tetraceros quadricornis*	*Antelope cervicapra cervicapra*
23.	Brown Goral (*Gural*)	*Nemorhoedus bubalina*	*Nemorhoedus goral goral*
24.	Himalayan Tahr(*Jharal*)	*Hemitragus jemlaicus*	*Hemitragus jemlahicus jemlahicus*
25.	Blue Sheep (*Bharal*)	*Ovis nahura*	*Pseudois nayaur nayaur*
26.	Tibetan antelope (*Pahari Mrig*)	*Kemas hodgsoni*	*Pantholops hodgsoni hodgsoni*
27.	Great Tibetan Sheep (*Nayan Argali*)	*Ovis ammon*	*Ovis ammon hodgsoni*
28.	Himalayan Ibex (*Tangrol*)	*Capra sibirica*	*Capra ibex sibirica*
29.	That/Serow (*Sarao*)	*Capricornis serow*	*Capricornis sumatraensis*
30.	Tibetan Gazelle	*Gazella tibeti*	*Procapra picticaudata picticaudata*
31.	Barking deer (*Kakar*)	*Cervulus aureus*	*Muntiacus muntjack*

The rapid decline of wildlife of Uttarkhand may be attributed to the following causes:

1. Demand of exploding population for agriculture/food;
2. Increased exploitation of forest products;
3. Clearing, felling of forest stands for timber and fuel;
4. Lack of breeding and rearing grounds due to their encroachment by man;
5. Poaching and illegal trade in hides, skins, furs, horns and antlers, etc.;
6. Heavy grazing by expanding population of cattle, sheep and goats, which nip off all emergent vegetation including planted forests;
7. Competition for food of our herbivorous domestic animals with herbivorous wildlife animals;
8. Shortage of herbivore animal food for the carnivorous wildlife.

The Palaearctic elements in Uttarkhand fauna are confined to high altitudes and, next to Kashmir, are much more pronounced here. Palaearctic elements have entered the Himalayas during the ice ages in the late tertiary period from Central Asia and Europe (Das 1966). The main Palaearctic species are the Urial, Tibetan Antelope, Tibetan Gazelle, Argali Sheep, Hoary Hill fox, Woolly wolf, Brown bear, Pallas' cat, Himalayan lynx, Snow leopard, Himalayan pinemarten, Siberian mink, Ermine, etc. These are the very species which are subjected to ecological and physical stress by Man, killing them mercilessly for their beautiful fur coats, hair, hides and horns.

It is, therefore, imperative to study the ecological and physiological adaptations of high altitude animals and to establish a string of sanctuaries, national parks and breeding centres for wildlife farming throughout the length and breadth of Uttarkhand Himalayas. This should be located at natural ecological altitudinal areas to suit the climatic and environmental need of the species to be conserved. It is sad that there are only four sanctuaries in the whole of Uttar Pradesh and only three in Uttarkhand, viz., Nanda Devi Sanctuary (325 sq. km.), G.B Pant Pashuvihar (1955 sq. km.) and the Corbett National Park (low altitude, 325 sq. km.). Ideally, small national parks that can be easily managed and also guarded should be established at three different altitudes in Uttarkhand, some at 4000 m., and others at 3000 m. and 2000 m. These would cover almost all the vanishing wildlife of Kumaun-Garhwal region. It may be pointed out that Himachal Pradesh has already established 24 Himalayan sanctuaries at different altitudes and Jammu and Kashmir at least five. There is no reason for Uttar Pradesh to lag so far behind.

REFERENCES

Atkinson, E.T. 1884. Fauna of the Himalayas. *Him. Gaz., I.*

Biswas, B. and Khajuria, H. 1955. Zoological results of the "Daily Mail" Himalayan Expedition, 1954. *Proc. Zool. Soc., Calcutta*, 9 : 229-253.

Biswas, B. and Tiwari, K.K. 1966. Taxonomy and distribution of the Indian Rodents. *Indian Rodent Symp., Calcutta.*

Bourliere, F. 1955. *Mammals of the World*. George Harrap, London.

Chakraborty, S. 1983. Contribution to the knowledge of the Mammalian fauna of Jammu and Kashmir State, India. *Rec. Zool. Surv. India, Occ. Paper* No. 38 : I-129.

Das, S.M. 1966. Palaearctic elements in the fauna of Kashmir. *Nature,* **212** : I - 6.

Ellerman, J.R. and Morrison-Scott, C.T.S. 1955. Check-list of Palaearctic and Indian mammals. *Brit. Mus. Nat. Hist., London*

Nath, Surendra 1993. A check list of Mammals of Jammu and Kashmir State (India) with remarks on the Mammal geography of the State. *Livestock Adv.,* 18(5) : 6 - 13.

Prater, S.H. 1965. *The Book of Indian Animals*. Bombay Nat. Hist. Soc.

Sanderson, I.T. 1955. *Living Mammals of the World*. Hamish Hamilton, London.

Walker, E.P. 1964. *Mammals of the World.,* John Hopkins Press, Baltimore.

Rec. Adv. Fish Ecol. Limn. Eco-conserv.
IV : 59-106

10

Studies on the Bioecology of Fishes of Jammu Province (Jammu and Kashmir State) India, Part IV : Food, Feeding Types and Morphological Adaptations of the Alimentary Canal of Some Fishes of Jammu Province

SURENDRA NATH

INTRODUCTION

The morphology of the organs of alimentation in fishes has attracted the attention of ichthyologists since very early times. The first indications of the relationships between the structure of the alimentary canal and the nature of feeding in fish are found in the voluminous works of Cuvier and Valenciennes (1828-1849), in which the authors observed that the ratio of intestinal length to body length was greater in herbivores than in carnivores, a fact that is true not only for fishes but also for many other groups of vertebrates. According to Cole (1944), Fabricius (1618) was the first anatomist to study the relationship of the length of the gut to the size of the body in several fishes, but he did not try to correlate it to their feeding habits. Jacobshagen (1911, 1913, 1915), working on a large number of fishes, agreed in general with the findings of Cuvier and Valenciennes (*op. cit.*), but at the same time he cited a number of exceptions to these findings. As the biology of fish, and in particular their feeding habits, were not being studied in detail at that time, only the qualitative aspect of feeding was examined and that too in a general way. There was no quantitative analysis of the significance of individual items of diet. Therefore, in most of the works published both then as well as later on (including the 1930s), the relationship between the structure of the alimentary canal and the nature of feeding was not studied, even in the monographs of Oppel (1898-1899) and Pernkof and Lehner (1937).

Detailed studies on the biology of feeding in fish facilitated the development of ecomorphological investigation of the alimentary system. This is evident in the works of Suyehiro (1942), Andriyashev (1944, 1946), Mookerjee and Das (1945), Rahimullah (1945), Al-Hussaini (1946, 1947, 1949, 1951), Hirsch (1950), Kapoor (1953), Das and Moitra (1955, 1956, 1958, 1963), and others. The interesting aspect of this problem is the study of the development of the alimentary system in fishes in relation to their feeding habits at different stages of ontogeny (Lange

1948), as well as the study of the influence of the food on the formation of the alimentary organs in ontogenetic development (Krayukhin 1959, Polyakov 1966).

Excellent historical reviews of early works on the functional anatomy of the alimentary canal in fishes have been published by Oppel (1896), Yung (1899), Sullivan (1907), Pictet (1909), Biedermann (1911) and Curry (1939). Outstanding works that have appeared since the publication of these reviews have been discussed at length by Das and Moitra (1955-1963) and, recently, by Kapoor, *et al.* (1975).

Some of the recent contributions in this field are: Rastogi (1966), Kamal (1966), Agrawal and Sharma (1966), Alexander (1966, 1967, 1969, 1970), Vukovic (1966), Keast and Webb (1966), Aganovic and Vukovic (1966), Bishop and Odense (1966), Branch (1966), Field (1966), Hickling (1966), Sehgal (1966), Singh (1966, 1967), Chitray (1966), Agrawal and Bala (1967), Belonozhko (1967), Yoshida (1967), Liem (1967), Maggese (1967), Moitra and Bhowmik (1967), Verighina (1967, 1969, 1971, 1977, 1979), Kamal (1967), Lal (1968), Shuljak (1968), Vasilieva and Korovina (1968), Keast (1968), Svetovidov and Skvorzowa (1968), Tandon and Goswami (1968), Verighina and Medani (1968), Vrba (1968), Bogachik (1969), Dalela (1969), Bertmar *et al.* (1969), Kamat (1969), Kariya (1969), Fukusho (1969), De Groot (1969, 1971), Ishiwata (1969), Khalilov (1969), Mehrotra and Khanna (1969), Schmitz and Baker (1969), De Martini (1969), Western (1969, 1971), Hill (1969), Agrawal and Tyagi (1969), Berry and Low (1970), Sehgal and Salaria (1970), Gaur and Jaish (1970), Kubota and Uyeno (1970) Loukashkin (1970), Nirmala and Sethi (1970), Khanna and Mehrotra (1970, 1971), Sriwastava (1970), Prasadam (1970), Swift (1970), Jearld and Brown (1971), Bucke (1971), Gupta (1971), Reutter (1971), Mester (1971, 1973), Das (1971), Pinkas et al (1971), Moitra and Sinha (1971), Taffanelli *et al.* (1971), Stroganov and Buzinova (1971), Pandey and Pandey (1972), Bremer (1972), Dewitt and Cailliet (1972), Dixit and Bisht (1972), Geistdorfer (1972), Alkholy and Malek (1972), Honma *et al.* (1972), Mikuriya (1972), Wilamowski (1972), Vegas-Velez (1972), Nelson (1972), Weisel (1973), Suvarova and Treschuk (1973), Korovina (1973), Gosline (1973), Mota *et al.* (1974), Nikolskaya and Verighina(1974), Ebeling and Cailliet (1974), Hobson (1974), Verighina and Savvaitova (1974), Verma and Dalela (1974), Verma and Tyagi (1974), Capanna *et al.* (1974), Hobson (1975), Sinha and Moitra (1975), Kafuku (1975), Mendelson (1975), Weinberg (1976), Adams (1976), Srivastava and Srivastava (1975), Sibbing (1976), Blaber (1976), Grizzle (1976), Witte and Barel (1976), Montgomery (1977), Stroband (1977), Moitra and Ray (1977, 1979), Chao and Musick (1977), Morse (1977), Kuwatani and Kafuku (1978), Pietsch (1978), Das and Pathani (1978), Chatterji et al (1978), Karamchandani and Mishra (1978), Kline (1978), Kumari and Nair (1979), Trewisan (1979), Aravindan (1980), Rao *et al.* (1980), Clarke and Witcomb (1980), Badola and Singh (1980), Saxena (1980), Marais (1980), Mok (1980), Singh *et al.* (1981), Singh and Samuel (1981, 1982), Biswas and Nassar (1981), Koslow (1981), Johar and Datta-Munshi (1981), Ghose *et al.* (1982), Verighina and Zholdasova (1982), Sharma and Tandon (1982), Datta and Das (1982), Das (1982), Hori (1983), Anthony (1983-1984), Sharma (1984), Parikh and Govindan (1984), Erondu (1984), Santhanakumar and Job (1983-1984), Datta and Dasgupta (1986), Biswas (1986), Bose and Islam (1986), Gupta

(1986), Nwadiaro and Okorie (1987), Sharma (1986-1937), Dasgupta (1988) and Channa and Trisal (1988).

OBSERVATIONS

The following eleven species of freshwater teleosts of Jammu province were studied during the course of present investigations for their food, feeding habits and morphological adaptations :

Sl.No.	*Family*	*Type*	*Feeding Habit*	*Locality*
1.	Cyprinidae	*Crossocheilus latius punjabensis* Mukerji	Herbivore	Poonch Valley
2.	Cyprinidae	*Labeo dero* (Hamilton)	Herbivore	Poonch Valley
3.	Cyprinidae	*Garra gotyla gotyla* (Gray)	Herbivore	Poonch Valley
4.	Cyprinidae	*Cyprinus carpio specularis* (Lacepede)	Herbivore	Jammu
5.	Cyprinidae	*C. carpio communis* Lin	Herbivore	Jammu
6.	Cyprinidae	*Rasbora (Rasbora) daniconius daniconius* (Hamilton)	Omnivore	Jammu
7.	Cyprinidae	*Barilius vagra vagra* (Hamilton)	Omnivore	Poonch Valley
8.	Cyprinidae	*Puntius conchonius* (Hamilton)	Omnivore	Jammu
9.	Cobitididae	*Noemacheilus rupecola rupecola* McClelland	Carnivore	Poonch Valley
10.	Cobitididae	*N. vittatus* (Heckel)	Carnivore	Poonch Valley
11.	Sisoridae	*Glyptothorax cavia* (Hamilton)	Carnivore	Poonch Valley

A. THE HERBIVOROUS FISHES

I. *Crossochielus latius punjabensis* Mukerji

Investigations by the present author on the food and feeding habits of *Crossocheilus latius punjabensis* obtained from Kashmir Valley and Poonch Valley (Jammu Province) have revealed that the fish is a predominantly benthophagous herbivore which feeds mainly on blue-green filamentous algae and higher acquatic plants. The animal component consists of aquatic insects, insect larvae and microcrustaceans. The plant food-animal food ratio was found to be 85 : 11, D.O.M., sand and mud forming about 5 per cent in the average annual feed of the fish.

The Oropharynx : The subterminal mouth is bordered by soft fleshy lips which are supported by edentulous jaws. The under surface of the jaw has a round, callous and rather elevated spot, which has a free anterior edge in the adult. The upper lip is fringed and deeply indented on the edge, while the lower lip is without any suctorial disc and is studded with fleshy papillae on its ventral surface. The lips are not continuous. The lower lip has a strong, sharp, horny covering. The position of the mouth and the character of the lips represent an intermediate condition between that of *Labeo dero* and *Garra gotyla gotyla* and indicate a benthophagous habit. A pair of rostral barbels is present. There is a U-shaped maxillary valve, with the limbs of the U-spread wide apart, so that it appears as a narrow arc of uniform thickness. The mandibular valve is of the same shape but is very much reduced.

Two median-patatal organs occur on the roof of the buccopharynx, one behind the other. The anterior palatal organ (dorsal lamellar organ) is roughly elliptical in outline, with a sharply-tapering posterior end; it bears a double row of fleshy papillae. A similar dorsal lamellar organ occurs on the roof of buccopharynx in *Labeo dero* but is absent in *Garra mullya* and *Garra gotyla gotyla*, despite the fact that both *Corssocheilus* and *Garra* are representatives of the same subfamily Garrinae (Das and Nath 1965, Nath 1978).

A well developed triangular palatal-pad lies behind the dorsal lamellar organ and is studded with small knob-like gustatory and respiratory papillae.

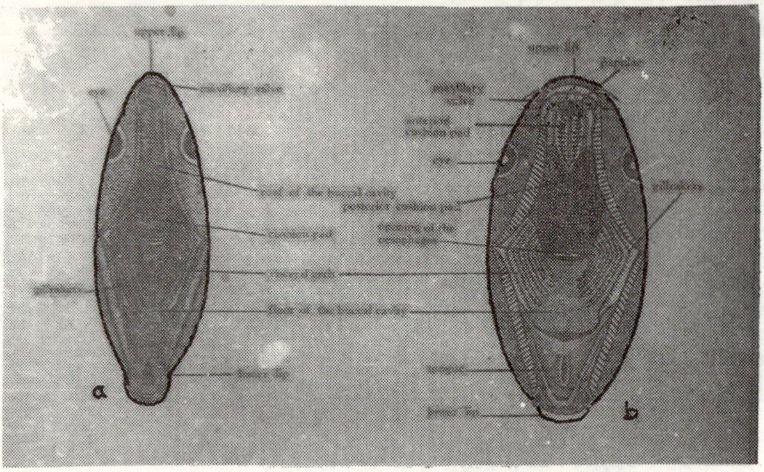

Plate I : a. Buccopharynx of *Rasbora daniconius*;
b. Buccopharynx of *Crossocheilus latius panjabensis*

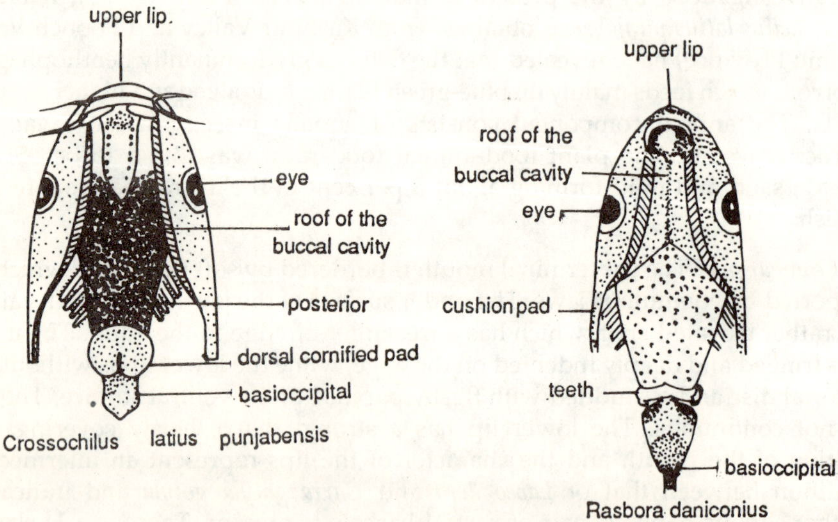

Plate IA : a. Roof of oropharynx of *Crossocheilus latius punjabensis*
b. Roof of oropharynx of *Rasbora daniconius*

The gill rakers are short, fine, soft and hair-like and are closely set on both sides of the branchial arches. The gill rakers of the adjacent arches interdigitate and thus form an efficient sieving mechanism.

A circular dorsal cornified pad occurs on the roof of the posterior pharynx, and is borne on the ventral surface of the masticatory process of the basioccipital. It has a distinct notch in the middle of its posterior margin. Small recurved inferior-pharyngeal teeth, borne on the ceratobranchials of the fifth pair of brachial arches, work against the dorsal cornified pad to provide a masticatory apparatus for crushing the aquatic plants ingested by the fish. The dental formula is 5, 4, 2/ 2, 4, 5.

The Gut : It is built on the typical herbivorous plan. A short oesophagus leads into the moderately thick walled intestinal bulb, the latter compensating for the absence of a true stomach. The intestine is very long and highly convoluted. The relative length of the gut ranges from 8.16 to 10.17 (average 9.14). A high relative length of the gut is shared by several other predominantly herbivorous teleosts such as *Labeo fimbriatus* (R.L.G. 22.0), *L. rohita* (R.L.G. 12.0), *L. kontius* (R.L.G. 11.0), *L. dero* (R.L.G. 8.86), *Garra gotyla gotyla* (R.L.G. 8.10) and *G. mullya* (R.L.G. 7.0). Hora (1951) attributed the extraordinarily high R.L.G. of teleosts such as *Labeo fimbriatus* and *L. kontius* to the probable presence of an excess of cellulose matter in their food, which naturally requires a relatively longer intestine for its proper assimilation.

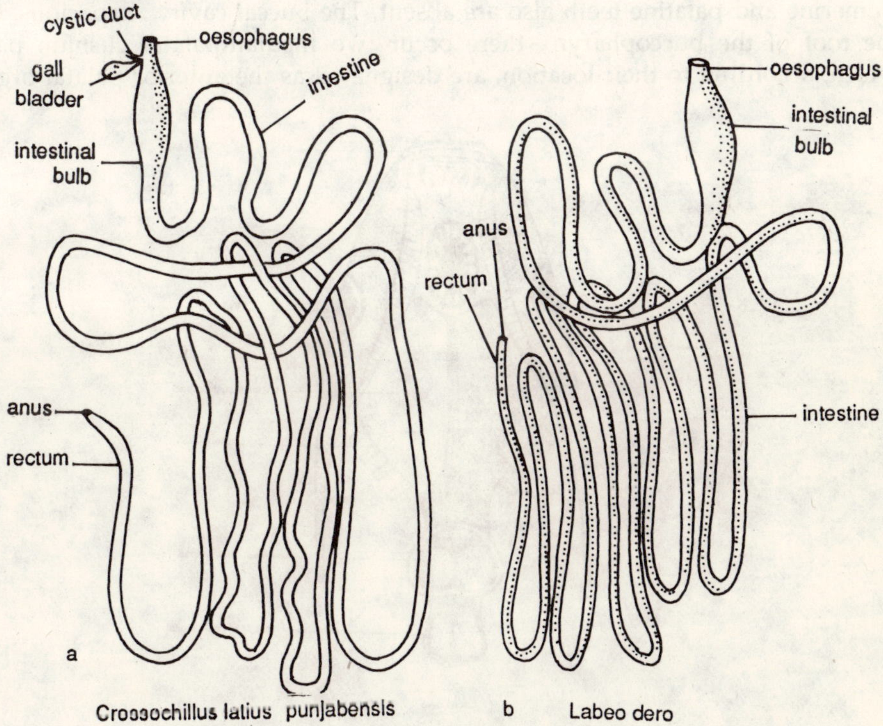

Plate II : a. Gut of *Crossocheilus latius panjabensis*; b. Gut of *Labeo dero*

Studies on the brain of *Crossocheils latius punjabensis* have revealed the presence of well developed optic and facial lobes, which indicates that *Crossocheilus* uses both sight and taste during feeding.

II. *Labeo dero* (Hamilton)

It is a completely herbivorous teleost which feeds chiefly at night, quietly grazing away at the tender leaves and shoots of aquatic plants. It resorts to plankton and mid-feeding even during the day. The sand grains often found in its gut contents suggest a partial benthophagous habit. The plant food component is about 80 per cent in the average annual feed, the animal component forming 15 - 20 per cent. Sand grains and allied inorganic detritus comprise 3 - 4 per cent. Seasonal fluctuations in the feeding pattern are marked, there being minimum feeding during December, January and February; the peak period is from April to June.

The Oropharynx : The mouth is rather narrow, transverse, crescentic and subterminal, bounded by non-protractile fleshy, fringed lips, which are continuous at the angles of the mouth, forming a labial fold and are lined internally by a cartilaginous base. The upper lip is soft, smooth and nearly semilunar; the lower lip is thicker and has a hard inner lining of horny nature, while its posterior edge is smooth, without any horny covering and is infringed with minute papillae. Both the jaws are edentulous. A single pair of short maxillary barbels is present. Vomerine and palatine teeth also are absent. The buccal cavity is spacious. On the roof of the buccopharynx there occur two median-palatal cushion pads which, according to their location, are designated as the anterior palatal organ

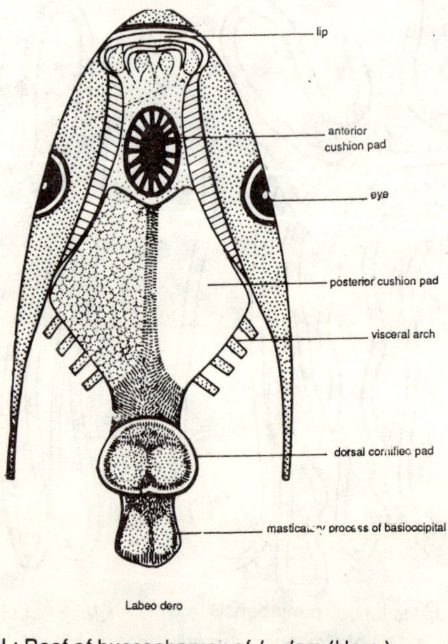

Plate III : Roof of buccopharynx of *L. dero* (Ham.)

(dorsal lamellar organ) and the posterior palatal pad (the palatal organ of Valatour). In a specimen 180 mm long, the anterior palatal organ was 6 mm long and about 3 mm in width, while the posterior palatal organ was 13 mm long and about 14 mm in width. The lamellar organ is elliptical and bears 10 - 13 fleshy comb-like processes which are extensions of the buccal mucosa; on either side of the lamellar organ occur two small "tentaculets" which also are fleshy and are backwardly directed. The posterior palatal pad is roughly hexagonal in outline and several times larger than the anterior palatal (lamellar) organ; it bears numerous papillae which are outgrowths from the mucous membrane lining the oropharyngeal cavity and are closely set. There is no tongue.

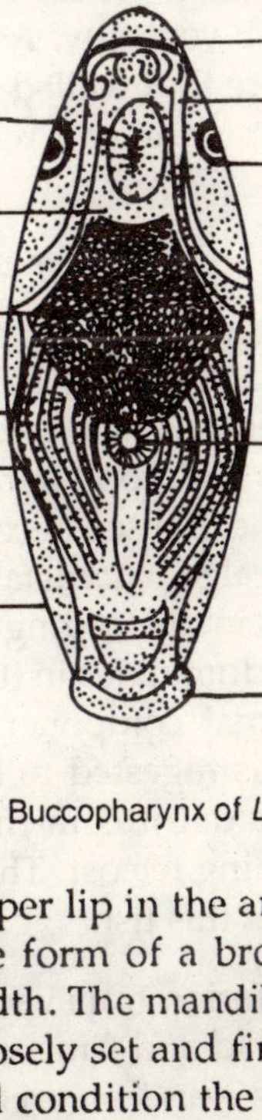

Plate IIIA : Buccopharynx of *Labeo dero*

Immediately behind the upper lip in the anterior part of the buccopharynx there is a maxillary valve in the form of a broad band of semicircular shape, which has an almost uniform width. The mandibular valve is absent.

The gill rakers are short, closely set and fine, almost hair like, about 1/3 as long as the eye; in the contracted condition the gill rakers of the adjacent arches interdigitate, forming a sieve for filtering out the phytoplankton from the outgoing water current.

The broad anterior pharynx narrows posteriorly to open into the posterior pharynx. On the roof of the latter there is present a nearly circular, thick, cornified pad (horny pad). It is hard and bears two circular depressions, and is borne on the masticatory process of the basioccipital. On the floor of the posterior pharynx occur strongly developed, prominent, hooked (plough-shaped) inferior pharyngeal teeth, arranged in three rows; the dental formula is 5, 4, 2/2, 4, 5. Each tooth is made up of a long shaft of dentine with a flat grooved enamelled molariform crown and a large pulp cavity. The inferior pharyngeal teeth are borne by the ceratobranchials of the fifth pair of branchial arches and work against the dorsal cornified pad to form an efficient grinding mechanism. The posterior pharynx has thick longitudinal mucosal folds which are continuous with the mucosal folds of the oesophagus.

The Gut : The posterior pharynx leads posteriorly into a short muscular tube, the oesophagus. The latter communicates posteriorly with a well developed intestinal bulb lying on the left side of the abdominal cavity. The intestinal bulb is marked internally by wavy longitudinal striations representing the mucosal folds. There is no stomach, the cystic duct from the gall bladder opening into the lumen of the intestinal bulb, which latter represents the hypertrophied anterior limb of the intestine.

The intestinal bulb leads into the intestine proper, which is a long, cylindrical, relatively thin-walled tube with numerous convolutions; the intestinal convolutions partially cover the intestinal bulb ventrally. As it approaches its termination, the intestine becomes more and more thin walled and leads into the rectum without any external demarcation. The pyloric and rectal caeca are absent. The relative length of the gut is about 8.86.

III. *Garra gotyla gotyla* (Gray)

Analyses of the gut contents of a large number of specimens of this Discognath collected from Poonch Valley have revealed that this teleost is a predominantly herbivorous teleost, which feeds on encrusting periphyton and blue-green filamentous algae, as both food items are abundantly available in the favourite haunts of the fish (i.e. expanses between the rapids). Small groups of this fish are seen rasping off the algal encrustations from the stones in the midst of streams. Seasonal fluctuations in the feeding of this Discognath have also been noticed, as the algal element predominates in its food during the months of algal bloom, i.e. April-May. The animal component in the average annual feed of *Garra* is about 2-5 per cent and is ingested to balance the diet. The presence of sand grains and allied inorganic detritus in the gut contents of *Garra* suggests that it is primarily a bottom-feeding teleost. This is also borne out by the salient features of the gut morphology of this fish.

The Oropharynx : The mouth is inferior, transverse, semicircular and overhung by the snout. It is bordered by a broad and fringed anterior (upper) lip in front - as in several other suctorial feeders - and an arched posterior (lower) lip behind. The lips are thick, fleshy and continuous, without any lateral lobes. The lower lip is lined internally by a cartilaginous cutting edge, which is used for scraping the encrusting periphyton off the surface of submerged rocks. A suctorial disc of semi-cartilaginous pad is present on the chin, formed on the lower lip. Two paris of barbels are present. The jaws are subequal and edentulous. The buccopharynx lacks vomerine and palatine teeth.

A well developed posterior median-palatal cushion pad (the palatal organ of Valatour), covered over with prominent and numerous respiratory papillae, occurs on the roof of the buccopharynx.

The gill rakers are few, slender, short and hair like and are closely set on both sides of the visceral arches. The posterior pharynx is dentulous. Well developed

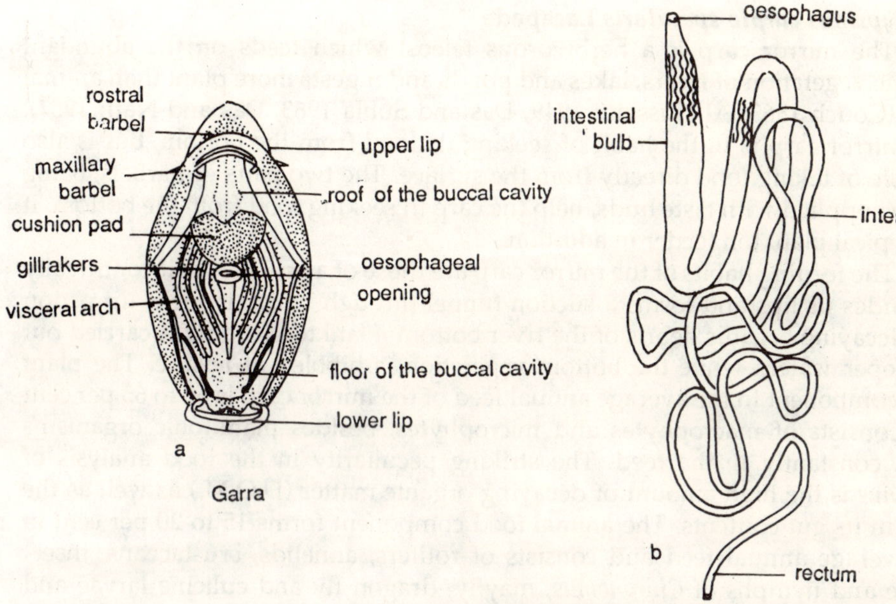

Plate IV : a. Oropharynx of *Garra gotyla* spread wide open
b. Gut of *Garra gotyla*, uncoiled

uncinate inferior—pharyngeal teeth, borne on the ceratobranchials of the fifth pair of visceral arches, work against the dorsal cornified pad present on the masticatory process of the basioccipital. The inferior pharyngeal teeth are arranged in three, closely approximating rows; the dental formula is 2, 4, 5/5, 4, 2.

The Gut : A short tubular oesophagus connects the pharynx posteriorly with the enlarged arm of the intestine—the intestinal bulb. A stomach in the true sense is absent, the large intestinal bulb partially compensating for the absence of a stomach by acting as a storage organ for the large amounts of aquatic plants and algal material that comprise the food of *Garra*. The cystic duct from the gall bladder opens into the lumen of the intestinal bulb immediately behind the junction of the oesophagus with the intestinal bulb. The wall of the intestinal bulb exhibits a pattern of a series of wavy longitudinal folds, which look like a series of Vs joined limb to limb, the Vs of the successive rows fitting into one another. These are the mucosal folds of the intestinal bulb which serve to increase the absorptive surface area of the intestinal wall.

Posteriorly the intestinal bulb narrows to lead into the rest of the intestine; the latter is coiled like a watch spring and covers the intestinal bulb ventrally. The surface of the whole of the intestine from the intestinal bulb to the rectum has a homogeneous appearance. The diameter of the intestine, with the exception of the intestinal bulb, is almost uniform; it decreases distinctly as the intestine approaches the rectum. Moreover, the wall of the intestinal bulb is relatively thicker than that of the rest of the intestine, while the rectum also has a distinctly thicker wall than the rest of the intestine. The relative length of the gut is about 8.19 mms.

IV. *Cyprinus carpio specularis* Lacepede

The mirror carp is a herbivorous teleost which feeds on the abundant aquatic vegetation of rivers, lakes and ponds and ingests more plant than animal food (Couch 1865, Al-Hussaini 1949, Das and Subla 1963, Das and Nath 1967). The mirror carp is in the habit of seeking its food from the bottom, but is also capable of taking food directly from the surface. The two pairs of harbels, being richly supplied with taste buds, help the carp in seeking food from the bottom. It is a typical plankton feeder in addition.

The feeding habits of the mirror carp are those of a grubber, as it continually protrudes its lips and forms a suction funnel through which it sucks up pedon and decaying organic debris of the river bottom. Plankton feeding is carried out in deeper waters while the bottom vegetation is nibbled at leisure. The plant food component in the average annual feed of the mirror carp is 60 to 65 per cent and consists of macrophytes and microphytes, besides planktonic organisms taken constantly in the feed. The striking peculiarity in the food analysis of *Cyprinus* is the high amount of decaying organic matter (D.O.M.) as well as the sand in its gut contents. The animal food component forms 15 to 20 per cent in the average annual feed and consists of rotifers, annelids, crustaceans, insect larvae and nymphs of *Chironomus*, mayfly, dragon fly and culicine larvae and molluscs. Although it is often stated that these animals are taken inadvertantly with the vegetation or with the D.O.M. at the lake bottom, yet, at the same time, the mirror carp has been seen to feed selectively, particularly on crustaceans and insect larvae. Another characteristic of the food of mirror carp is the large amount of organic debris which it ingests by suctorial feeding, particularly during the high feeding periods. Seasonal fluctuations in the annual food cycle have also been observed in relation to the prespawning and postspawning periods of the fish.

The Oropharynx : The mouth is terminal, oblique and depressed, its cleft not extending to the anterior margin of the eyes. It is provided with soft, fleshy and unfringed semicircular lips and supported by edentulous jaws. The upper jaw is more or less projecting. The maxillary is protractile, reaching to anterior nostril. One pair each of maxillary and rostral barbels is present.

The protrusible mouth opens into a depressed oropharyngeal cavity, which is narrow anteriorly but widens posteriorly. Immediately inside the upper lip there is present a crescent-like maxillary valve; the mandibular valve is quite reduced, being almost vestigial. On the roof of the oropharynx there is present a thick posterior median-palatal cushion pad (the palatal organ of Valatour) which covers the greater part of the roof. It is roughly triangular in outline, being narrow anteriorly and broad posteriorly. In a specimen 108 mm long, the palatal pad was 12 mm long (antero-posteriorly) and 14 mm broad (from side to side) at the widest region. The knob-like papillae of the buccopharynx continue over the surface of the cushion pad.

The floor of the buccopharynx is supported by the basihyal and lined by a thick mucous membrane, which is produced into minute papilla-like protuberances; these papillae are gustatory in function. On the floor of the

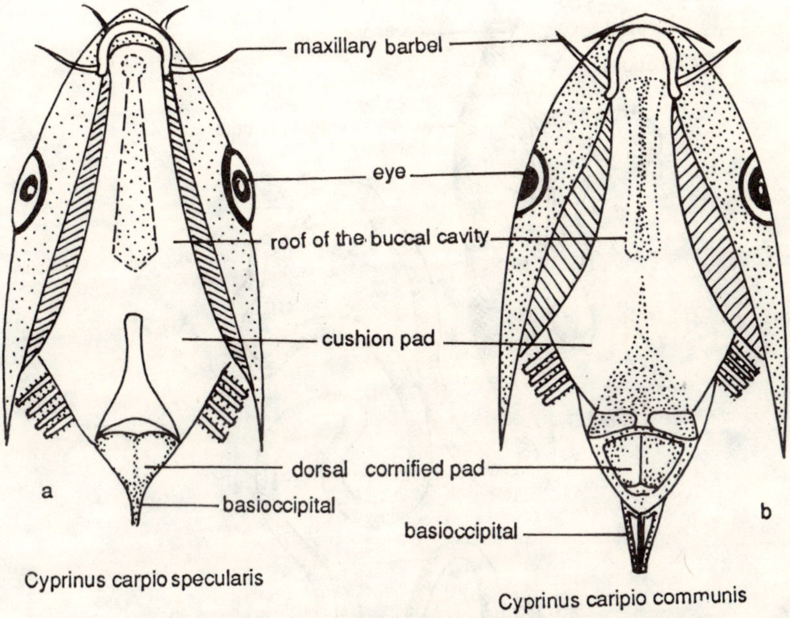

Plate V : a. Roof of Oropharynx of *Cyprinus carpio specularis*
b. Roof of Oropharynx of *Cyprinus carpio communis*

buccopharynx lies the rudimentary tongue. Longitudinal rows of papillae are found on the anterior part of this structure.

The gill rakers are fairly long and hair like, measuring about 1 mm in length in a fish 108 mm long. They are closely set on both sides of the branchial arches. The gill rakers of the adjacent arches nearly interdigitate as in the planktophagous fishes. The long gill rakers are characteristic of the majority of benthophages which stir up the mud—a habit also possessed by the mirror carp (Kyle 1926).

The posterior pharynx is considerably narrower and masticatory in function. A hard, nearly triangular callous pad (the dorsal cornified pad) is borne by the basioccipital on its ventral side and is about 5 mm long and 6 mm in width. This pad lies posterior to the median palatal cushion pad, with its base directed anteriorly, and has been designated as the "dorsal cornified pad" by Curry (1939). Several prominent molariform teeth are borne by the ceratobranchials of the fifth pair of branchial arches and are known as the inferior-pharyngeal teeth. These work against the dorsal cornified pad and form an efficient masticatory apparatus for crushing the hard shells of molluscs ingested by the fish.

The Gut: The posterior pharynx leads posteriorly into a narrow, tubular oesophagus, which has thick, muscular walls. On entering the body cavity the oesophagus opens into an elongated sac-like intestinal bulb without the intervention of a stomach. The absence of the stomach is partially compensated by the enlargement of the first limb of the intestine, which has thick, elastic walls and is known as the intestinal bulb. It has a tapering form, being wide in the middle and relatively narrow at the ends, and lies dorsal to the intestinal

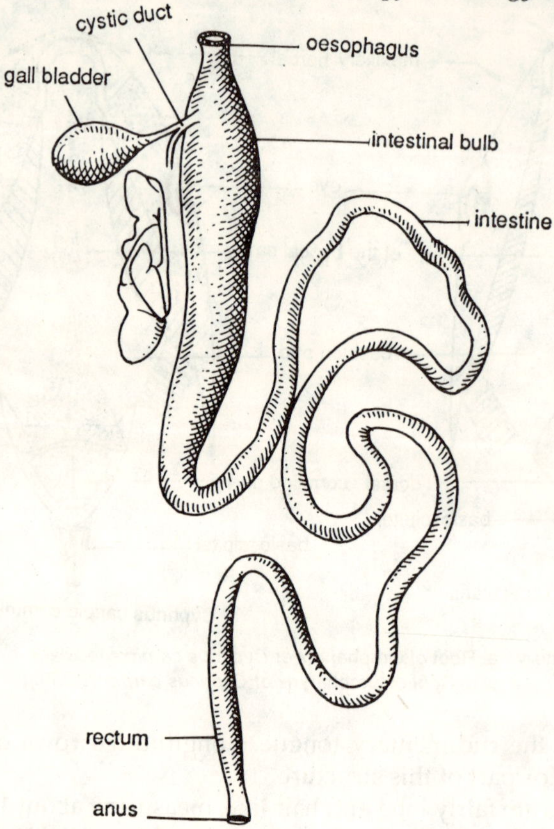

Plate VI : Gut of *Cyprinus carpio specularis* uncoiled.

convolutions. A little below the passing of the oesophagus into the intestinal bulb a short duct, the cystic duct from the gall bladder, opens into the lumen of the bulb. The posterior extremity of the bulb curves to the left and passes into the comparatively thin-walled portion of the intestine, which has an almost unifrom diameter. The intestinal convolutions are interconnected by folds of mesentery, to which are attached groups of fat cells. Posteriorly, the intestine widens slightly to form the rectum which is thin walled. The relative length of the gut is about 2.00mms.

V. *Cyprinus carpio communis* Linn.

Like the mirror carp, the scale carp also is herbivorous and feeds on the soft aquatic vegetation. It may also ingest some benthonic invertebrates occasionally. The scale carp is a voracious herbivore, feeding on coarse macrophytes and insect adults, besides plankton algae, zooplankton (rotifers) and decomposed organic material (D.O.M.) which it ingests in large quantities by sucking from the bottom of the water body. It thus shows little specificity in its food. The nature of its diet varies from season to season. The plant food-animal food ratio in the average annual feed of the scale carp is 57 : 34, the rest (D.O.M.) making up about 9 per cent. The young feed for a certain period of time on zooplankton, rotifers, crustaceans, etc. but when they reach about 18 mm in length, they begin to feed on benthonic invertebrates. The adult fish feeds primarily on benthonic invertebrates, such as ostracods, molluscs, the chironomid larvae and other bethonic animals. It also feeds on vegetation to a very considerable extent in lakes and reservoirs which are heavily overgrown with vegetation.

The Oropharynx : Mouth is terminal, oblique, its cleft not extending to the anterior margin of the eyes. The depressed oropharyngeal cavity is bounded anteriorly by the soft fleshy and unfringed lips supported by edentulous jaws. The upper jaw is more or less protractile, reaching to the anterior nostril. A pair each of maxillary and rostral barbels is present. A crescentic maxillary valve is present; its surface is beset with small knob-like papillae. The mandibular valve is vestigial. The whole of the oropharynx is lined internally with a single soft mucous membrane which is produced into mucous-secreting gustatory papillae. Immediately behind the upper lip there lies a flap of tissue which is attached in front but free behind. A median-palatal cushion pad covers the greater part of the roof; it is roughly triangular, being narrow anteriorly and broad posteriorly. In a specimen 250 mm long, this pad was 26 mm in length (anteroposteriorly) and 21 mm in (maximum) width. The knob-like papillae of the buccopharynx continue over the surface of the cushion pad and are much more prominent in this region. On the floor of the buccopharynx lies the rudimentary tongue.

The gill rakers are fairly long and soft, measuring 2-4 mm in length and are closely set on both sides of the branchial arches. As in the mirror carp, the gill rakers of the adjacent arches nearly interdigitate, forming a sieving mechanism for the plankton.

The posterior pharynx is a triangular chamber, which is broad anteriorly and narrow posteriorly. The dorsal cornified pad is roughly triangular, 8 mm in length and 11 mm in width in a fish 250 mm long and is borne on the masticatory process of the basioccipital. Its apex is directed posteriorly while the broad base is directed anteriorly. Several strong molariform inferior-pharyngeal teeth are present on the ceratobranchials of the firth pair of branchial arches; by working against the dorsal cornified pad, they form an efficient masticatory apparatus.

The Gut: It comprises a narrow muscular oesophagus which passes imperceptibly into a large, thick walled, tubular structure—the intestinal bulb. A true stomach is absent in the scale carp (as well as in the mirror carp). The absence of the stomach is partially compensated by the hypertrophy of the first limb of the intestine to form the large intestinal bulb. The bulb has a tapering form, being widest in the middle and relatively narrow at the ends; it lies dorsal to the intestinal comvolutions. In a specimen 250 mm long, the length of the intestinal bulb was 68 mm.

A little below the merger of the oesophagus with the intestinal bulb, a stout cystic duct from the gall bladder opens into the lumen of the bulb. Posteriorly the bulb narrows distinctly and passes into the main mass of the intestine which has relatively thin walls and an almost uniform diameter. The intestinal

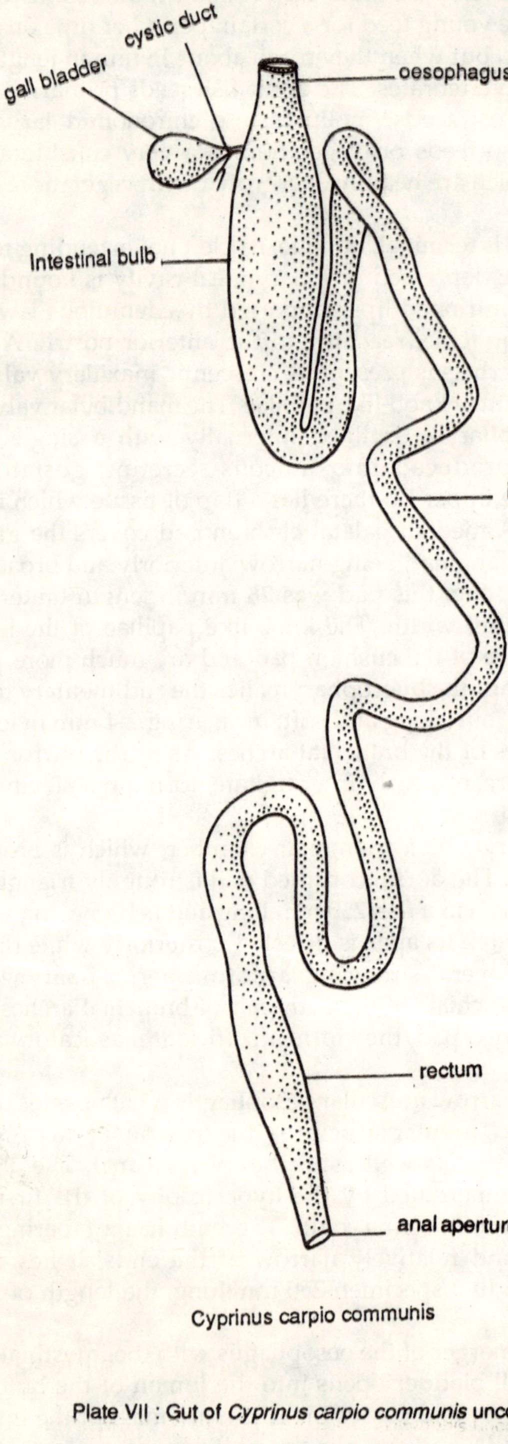

Plate VII : Gut of *Cyprinus carpio communis* uncoiled.

convolutions are interconnected by prominent mesenteries, with groups of fat cells attached to them.

The intestine widens slightly at its posterior end to merge with the rectum. The relative length of the gut is about 1.92, being thus slightly lesser than that of the mirror carp.

B. THE OMNIVOROUS FISHES

VI. *Rasbora (Rasbora) daniconius daniconius* (Hamilton)

Rasbora daniconius is an omnivorous teleost which feeds on crustaceans, aquatic annelids and soft aquatic plants, the animal components predominating in the average annual feed. The fry ingest large quantities of infusoria, rotifers and microworms. The animal food-plant food ratio has been found to be 65 : 35 in the average annual feed of the fish, which can, therefore, be designated as a carniomnivore (Das and Moitra 1963). Seasonal fluctuations in the animal food-plant food ratio have also been observed.

The Oropharynx: The mouth is oblique and moderately wide, with a slightly protruding lower jaw. The rather uppishly directed mouth indicates the capacity of the fish to obtain food from the surface. The lips are thin, soft and unfringed and supported by edentulous jaws. The lower lip is simple. The lower jaw bears one central and two internal prominences, one on each side, which fit into corresponding emrginations on the upper jaw. Barbels are absent. Mouth leads into a depressed buccopharynx. A membranous semicircular maxillary valve and a vestigial mandibular valve are present. The mucous membrane lining the buccopharyngeal cavity is produced into minute papillae which are respiratory as well as gustatory in function. On the posterior portion of the roof of the buccopharynx there lies a well developed median-palatal cushion pad with its base directed anteriorly. The surface of the cushion pad is also covered with the mucous membrane and produced into relatively larger respiratory papillae. In a specimen 55 mm long, the cushion pad was 4 mm in length and 3 mm in width. There is no distinct tongue.

The gill rakers are closely set, short and lanceolate, measuring 0.5 mm in average length and are borne on both sides of the branchial arches. The gill rakers of the adjacent arches interdigitate, so that sieving of the outgoing water current is possible in this fish.

Posterior to the median palatal cushion pad and borne on the ventral surface of the basioccipital there is present a poorly developed, dorsal cornified pad, which is roughly triangular, with its apex directed posteriorly. It is about 3 mm long and 2 mm in width at the base. Small, sharp and curved inferior-pharyngeal teeth are borne by the ceratobranchials of the fifth pair of branchial arches, which work against the dorsal cornified pad and provide a simple masticatory apparatus. The dental formula is 5, 3, 2/2, 3, 5.

The Gut: The gut is built on the same basic plan as in the herbivorous teleosts, but is relatively shorter. A short and fairly muscular oesophagus directly leads into an enlarged intestinal bulb which represents a functional dilatation of the anterior

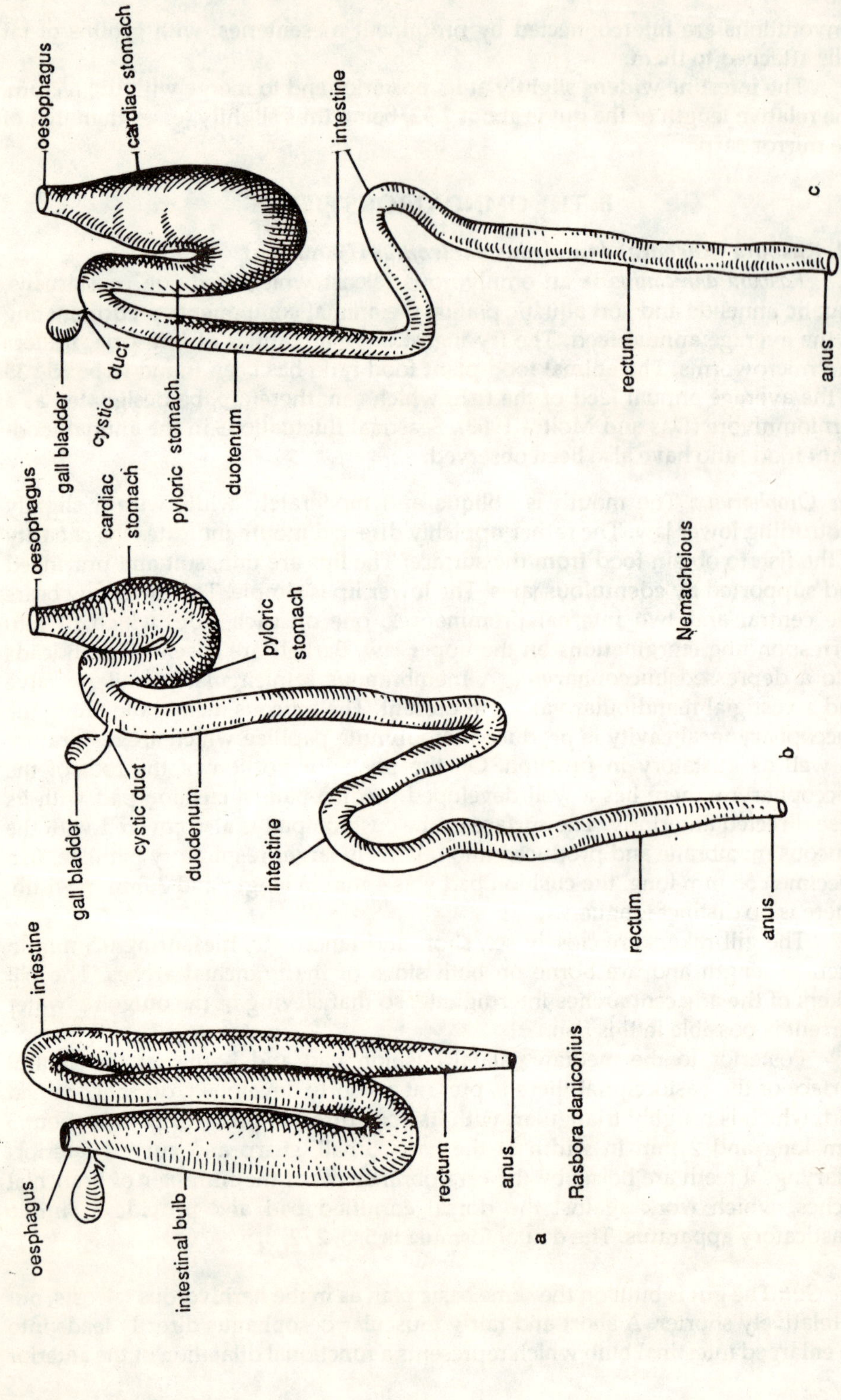

Plate VIII : a. Gut of *Rasbora daniconius* uncoiled; b. Gut of *Nemacheilus vittatus* uncoiled c. Gut of *N. rupicola* uncoiled

limb of the intestine, serving as a reservoir in which storage and digestion of the ingested food occurs. A true stomach is absent. The cystic duct opens into the intestinal bulb immediately behind the junction of the oesophagus with the bulb. The rest of the intestine has thin walls and an almost uniform diameter throughout its length. It is arranged in the form of two limbs which run parallel to the intestinal bulb. Pyloric and rectal caeca are absent. The relative length of the gut is about 1.10mms. Being an omnivorous teleost, *Rasbora* has a gut almost equal in length to its total body length, representing an intermediate stage between that of a true herbivore and a carnivore.

VII. *Barilius vagra vagra* (Hamilton)

Like several other small-bodied hill stream cyprinids, such as *Rosbora daniconius* and *Puntius conchonius, Barilius vagra vagra* is an omnivorous teleost which feeds on aquatic annelids, micro-crustaceans, rotifers and soft aquatic plants. The animal components predominate in the average annual feed of the teleost. The animal food-plant food ratio in the average annual feed of the fish was found to be 70 : 30, the fish being, thus, a carni-omnivore. Seasonal fluctuations in the feeding of the teleost have also been noticed, the insect component being much more predominant during the months of May, June and July. *Barilius* inhabits the upper water levels and has been observed to take flying insects from the surface of water.

The Oropharynx: The mouth is nonprotractile, anterior, somewhat oblique, having a moderately deep cleft. It is bounded by soft, thin and unfringed lips which are supported by edentulous jaws. The jaws are compressed, the upper jaw longer than the lower; the lower jaw has a knob above the symphysis. The upper jaw has a corresponding emargination to receive the knob. This is a characteristic feature of predatory carps such as *Aspius aspius, Barilius moore* and *B. christy* (Kazansky 1964, Matthes 1963). The posterior extremity of maxilla extends to beneath the middle of the orbit. The jaws are almost equal in length anteriorly, the upper slightly notched and the lower slightly larger and covered with large pores in the adult. There are two pairs of barbels, the rostral barbels being longer while the maxillary barbels are very short.

The mouth leads posteriorly into a depressed oropharyngeal cavity. The maxillary valve is semicircular and better developed than the mandibular valve; the latter is crescentic and encricles the anterior part of the base of the 'tongue'. The mucous membrane lining the oropharyngeal cavity is produced into numerous gustatory papillae which are arranged in rows converging towards the mid-dorsal line. The posterior two-thirds of the roof of the oropharynx bear a median-palatal pad, which is roughly triangular in outline. The surface of the palatal-pad is also covered by rows of gustatory papillae arranged in irregular rows. In a specimen 50 mm long, the palatal pad was 4.59 mm long (antero-posteriorly) and 2.0 mm in maximum width.

There is no distinct tongue. It is, however, represented by a conspicuous thickening of the floor of the oropharynx, which becomes much more prominent posteriorly near the origin of the visceral arches. The "tongue" is also covered

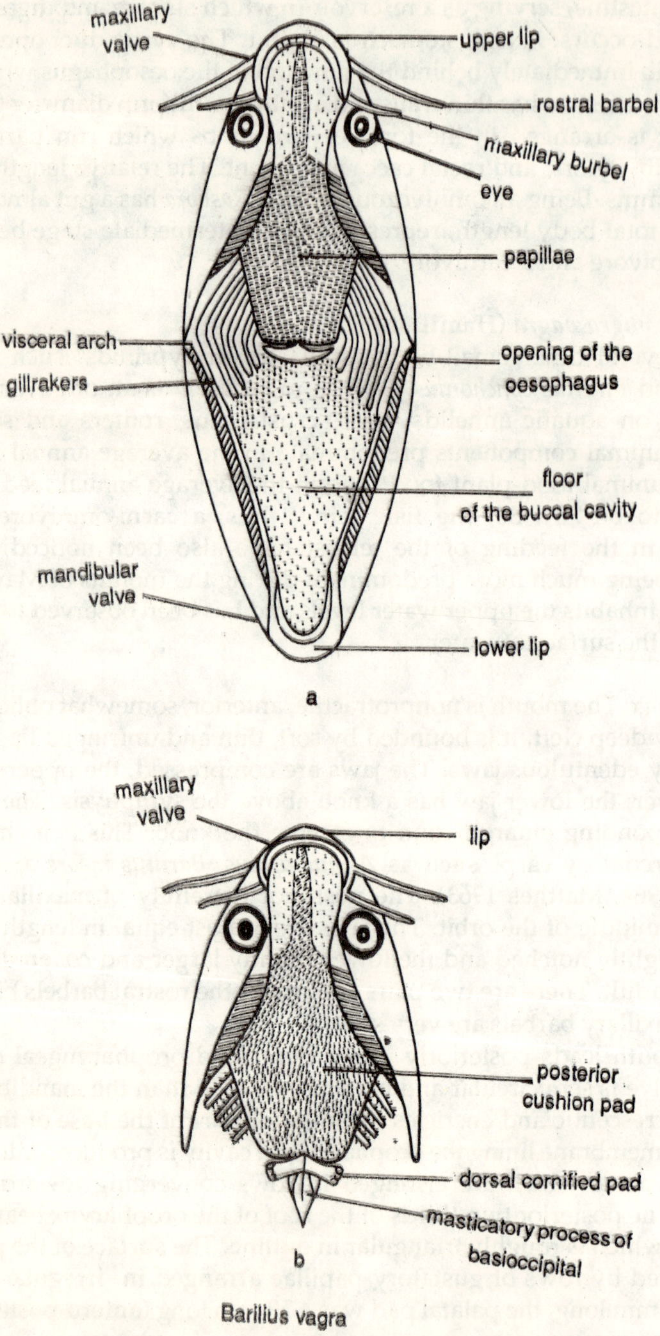

Plate IX : a. Oropharynx of *Barilius vagra* spread wide open; b. Roof of Oropharynx of *Barilius vagra*

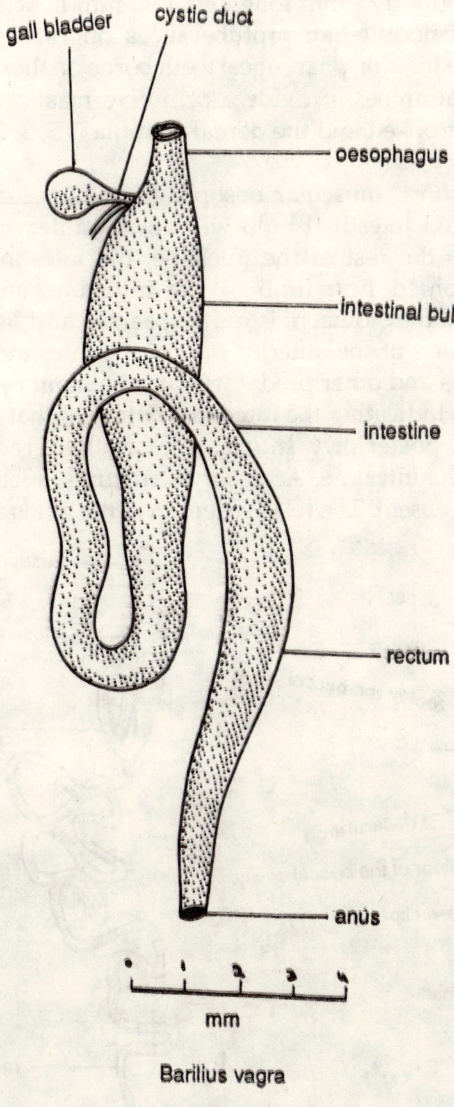

Plate X : Gut of *Barilius vagra* uncoiled

with the minute gustatory papillae, but it does not become sufficiently freed from the floor of the oropharynx so as to form a definite organ entity.

The gill rakers are very short, soft and small structures, measuring about 0.5 mm in average length and are closely set on both sides of the visceral arches. The gill rakers of the adjacent arches nearly interdigitate.

Posterior to the palatal pad the oropharyngeal cavity narrows considerably. On the ventral surface of the basioccipital there is borne a ^ shaped dorsal cornified pad, which is about 0.75 mm long and 1.50 mm in width. It is poorly developed and bears 2-3 nodule-like protuberances on each side which, by working against the minute inferior-pharyngeal teeth borne on the ceratobranchials of the fifth pair of visceral arches, provide a primitive masticatory apparatus. The pharyngeal teeth are crooked and the dental formula is 5, 3, 2/2, 3, 5.

The Gut: It comprises a short, muscular oesophagus which leads into a well developed and thick walled intestinal bulb without the intervention of a true stomach. As compared to the rest of the intestine, the intestinal bulb, which represents the hypertrophied first limb of the intestine, and is a typical morphological analogue of the stomach, is quite spacious and lies dorsal to the intestinal loop. The widening of the anterior part of the intestine is much more clearly expressed in *Barilius* and other predatory carps. A stout cystic duct opens into the lumen of the bulb, indicating the intestinal nature of that structure.

The intestine leads posteriorly into a thick walled rectum which is considerably wider than the intestine. Accessory structures, such as the pyloric and rectal caeca, are totally absent. The relative length of the gut is about 0.56mms.

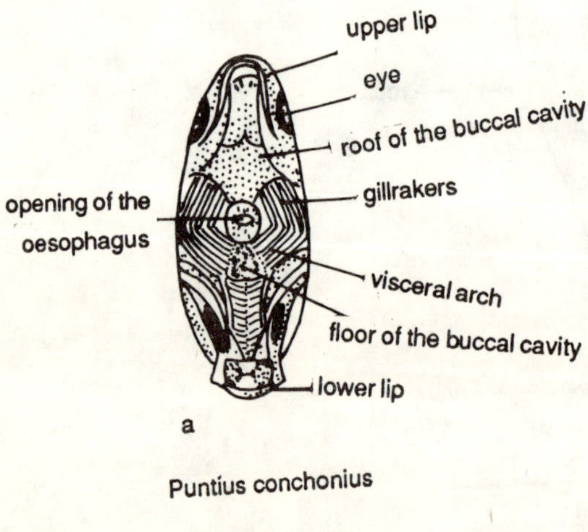

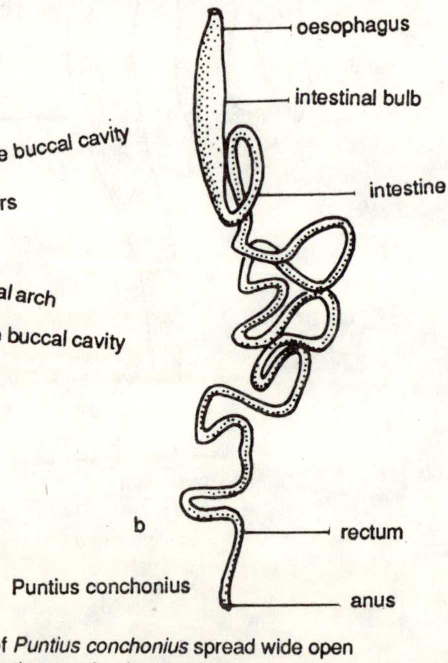

Plate XI : a. Oropharynx of *Puntius conchonius* spread wide open
b. Gut of *Puntius conchonius* uncoiled

VIII. *Puntius conchonius* (Hamilton)

Puntius conchonius feeds on algae, higher aquatic plants, bryozoans, insects and micro-crustaceans. Plant organisms, however, are by far the predominant elements in its diet, comprising about 58 per cent in the average annual feed of the fish. The animal food-plant food ratio in the average annual feed is about 30 : 58, the fish being, therefore, a herbi-omnivore.

The Oropharynx: The mouth is arched, anterior, nonprotrusible and bounded by soft, thin and unfringed lips which are almost crescentic in shape. The lips cover the jaws and lack a symphysial knob. Barbels are absent. Posteriorly the mouth leads into a depressed oropharyngeal cavity. The mucous membrane lining the oropharynx is produced into minute papillae at places.

The gill rakers are produced into thorn-like projections; the rakers of the adjacent visceral arches interdigitate to form a sieves like structure for filtering out plankton organisms. There are no palatine or vomerine teeth. A dorsal cornified pad is present on the masticatory process of the basioccipital. The inferior-pharyngeal teeth are borne on the ceratobranchials of the fifth pair of branchial arches and work against the cornified pad to form a simple masticatory apparatus.

The Gut: The pharynx leads posteriorly into a short and rather muscular oesophagus which, in turn, communicates with the enlarged intestinal bulb. The bulb has relatively thinner walls than the oesophagus. Immediately behind the junction of the oesophagus with the intestinal bulb, the cystic duct from the gall bladder opens into the lumen of the bulb. The rest of the intestine (behind the bulb) is also thin walled and has an almost uniform diameter throughout its length. It is thrown into a few (5-6) coils which overlie the intestinal bulb ventrally. The intestine leads into the rectum which has thicker walls than the intestine. The relative length of the intestine is about 3.25-3.27mms.

C. THE CARNIVOROUS FISHES

IX. *Noemacheilus rupecola rupecola* (McCelland)

It is a carnivorous cobitid loach which feeds mainly on benthonic invertebrates such as the chironomid larvae, aquatic annelids, crustaceans, bottom molluscs, fish eggs, etc., along with a small amount of soft aquatic vegetation such as the blue-green algae and encrusting periphyton. The adult insects in the food of this cobitid are *Corixa* as well as the small adult insects which fall into the water, such as butterflies, dragonflies, etc. The plant food component in the average annual feed is less than 10 per cent and consists of microphyton attached to the stones under which this loach lives. The plants ingested are filamentous algae and some diatoms. Being a bottom-feeder, it ingests a small amount of sand and allied inorganic detritus along with its food, up to the extent of 5-10 per cent.

Seasonal fluctuations in the annual feed of *N. rupecola* are considerable. During the months of algal-bloom this loach exhibits a steep rise in the intake of plant food. During the post-spawning season it takes up less than half of its

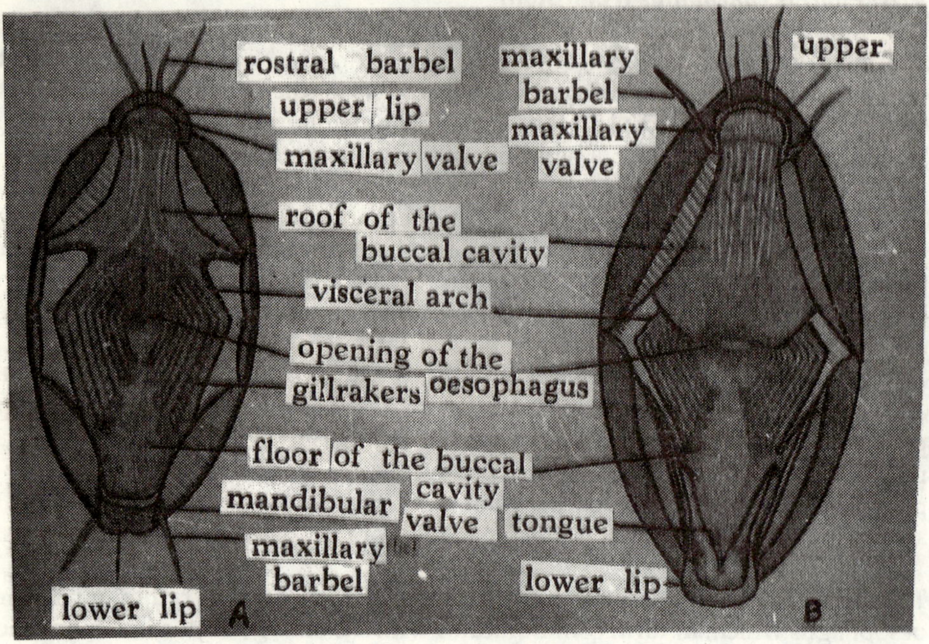

Plate XII : a. Oropharynx of *Nemacheilus rupecola* spread open
b. Oropharynx of *N. vittatus* spread open

usual food intake. There is a steep rise in feeding from February to April. Similar observations were recorded by Das and Nath (1967) for *N. kashmirensis* Hora, a species endemic to Kashmir Valley.

N. rupecola possesses small, reduced eyes and, because of the free-flowing sand and silt shooting past its head, it does not feed by sight. Therefore, food location is mainly the responsibility of sensory organs located on the barbels. Singh and Khanna (1978) showed an interesting correlation between the brain structure and habits of *N. rupecola*. As the loach gropes for food on the bottom with the help of its barbels—innervated by fibres from the facial nerve—the facila lobe of the medulla is highly developed. The large valvula cerebelli and the well developed posterior mesencephalo-cerebellar and the lobo-cerebellar tracts of the cerebellum are correlated with the active habits and well developed gustatory sense of the fish respectively. The sense of vision plays little role in its life activities and hence the fish has small optic lobes.

The Oropharynx: The mouth is small, inferior, transverse and subterminal, with thick, papillated lips which assist in temporary attachment to submerged stones. The upper and lower lips are confluent at the angles of the mouth. The lower lip is interrupted in the middle and the anterior (upper) lip is not hypertrophied. The jaws are edentulous. Mouth opens into a depressed oropharyngeal cavity which is narrow anteriorly but widens posteriorly. The maxillary, mandibular, palatine and vomerine teeth are totally absent. The maxillary and mandibular

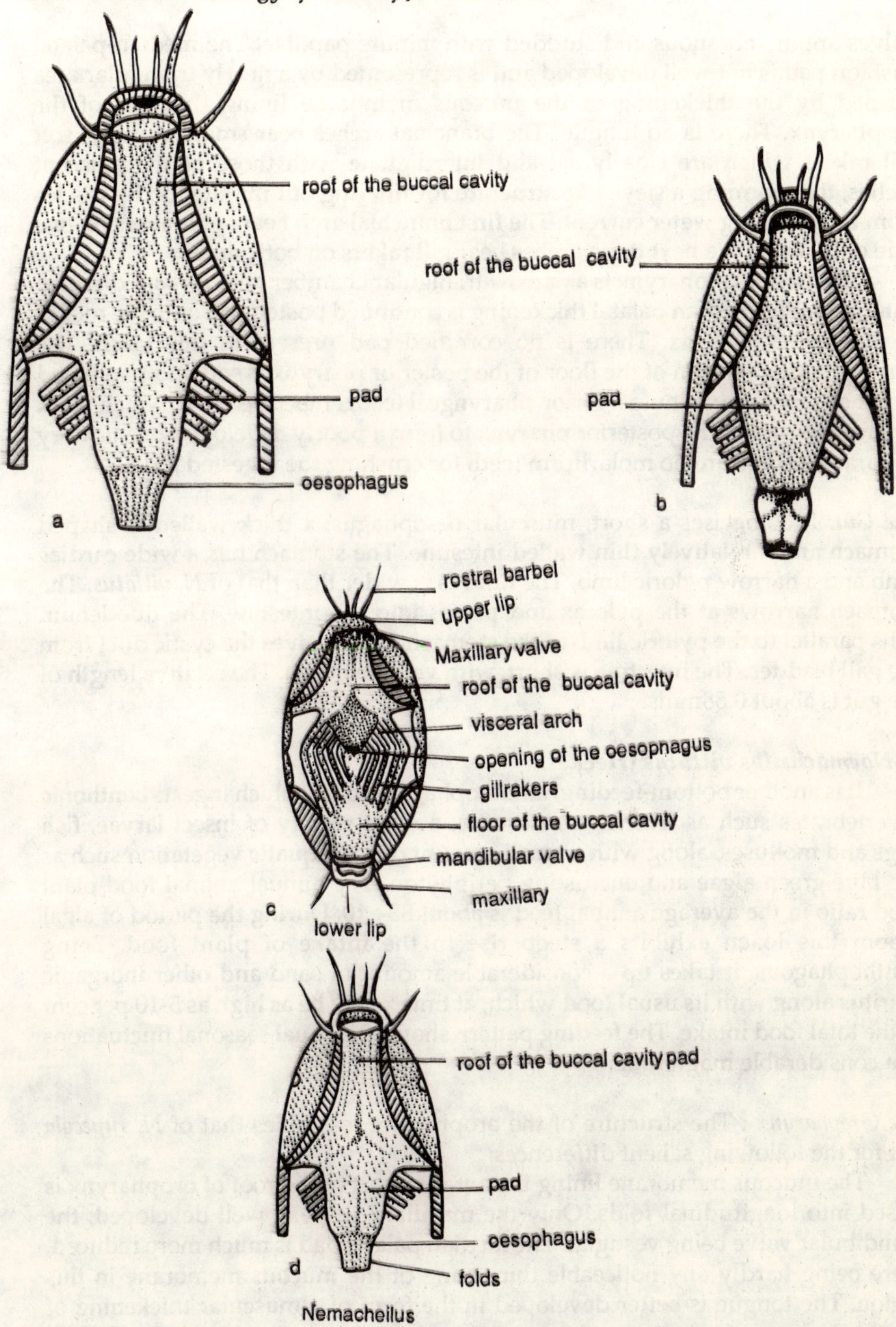

Plate XIII : a. Roof of Oropharynx of *Nemacheilus rupecola*; b. Roof of Oropharynx of *N. vittatus*; c. Oropharyx of *N. rupeola* spread open; d. Roof of Oropharynx of *N. rupecola*

valves are membranous and studded with minute papillae. The median-palatal cushion pad is not well developed and is represented by a nearly triangular area formed by the thickening of the mucous membrane lining the roof of the oropharynx. There is no tongue. The branchial arches bear small, stumpy, soft gill rakers which are closely set and interdigitate with those of the adjacent arches, thus forming a sieve like structure for filtering out minute food particles from the outgoing water current. The first branchial arch bears gill rakers on one side only, while the next three arches bear gill rakers on both sides.

The posterior pharynx is a narrow triangular chamber with its base directed anteriorly. The median palatal thickening is continued posteriorly onto the roof of the posterior pharynx. There is no cornified pad present in this loach. The corresponding portion of the floor of the posterior pharynx is soft and papillated and there are a few minute inferior-pharyngeal teeth embedded in it, which work against the roof of the posterior pharynx to form a poorly developed masticatory apparatus. There are no molariform teeth for crushing the ingested food.

The Gut: It comprises a short, muscular oesophagus, a thick walled U-shaped stomach and a relatively thin walled intestine. The stomach has a wide cardiac limb and a narrow pyloric limb. The fundus is wider than that of *N. vittatus*. The stomach narrows at the pyloras and passes into the intestine. The duodenum runs parallel to the pyloric limb of the stomach and receives the cystic duct from the gall bladder. The intestine is short, with very few coils. The relative length of the gut is about 0.88mms.

X. *Noemacheilus vittatus* (Heckel)

It is another bottom-feeding, entomophagous loach which ingests benthonic invertebrates such as worms, crustaceans, a wide variety of insect larvae, fish eggs and molluscs, along with a small amount of soft aquatic vegetation such as the blue-green algae and encrusting periphyton. The annual animal food-plant food ratio in the average annual feed is about 85 : 10. During the period of algal bloom this loach exhibits a steep rise in the intake of plant food. Being benthophagous, it takes up a considerable amount of sand and other inorganic detritus along with its usual food which, at times, may be as high as 5-10 per cent of the total food intake. The feeding pattern shows the usual seasonal fluctuations of a considerable magnitude.

The Oropharynx : The structure of the oropharynx resembles that of *N. rupecola* but for the following salient differences:

The mucous membrane lining the anterior part of the roof of oropharynx is raised into longitudinal folds. Only the maxillary valve is well developed, the mandibular valve being vestigial. The median palatal pad is much more reduced, there being hardly any noticeable thickening of the mucous membrane in this region. The tongue is better developed in the form of a muscular thickening of the floor of the anterior part of the oropharynx and is noticeably elevated from the floor; however, it is not detached from the floor of the oropharynx so as to form a definite organ entity. The lower lip is entire. The gill rakers are slender and spine like.

The posterior pharynx is a narrow chamber in which there is no dorsal cornified pad. The median palatal cushion pad (the muscular pad of Khanna and Pant 1964) extends posteriorly into the cavity of the posterior pharynx; the corresponding part of the floor of the posterior pharynx is soft and papillated. The inferior-pharyngeal teeth are few and minute and are embedded in the floor of the posterior pharynx; by working against the dorsal continuation of the muscular pad they form a sort of primitive masticatory apparatus.

The Gut: It is built on the typical carnivorous plan. The muscular oesophagus leads into a pitcher-like stomach which is a muscular sac with a wide fundus. There is no external demarcation at the point of junction of the oesophagus with the stomach. The latter is J-shaped, with the cardiac and pyloric limbs nearly parallel to one another. There are no pyloric or rectal caeca. The pyloric end of the stomach joins with the duodenum, which receives the cystic duct from the gall bladder. A distinct groove on the outside demarcates the junction of the pyloric end of the stomach with the duodenum. The latter has thinner walls than the stomach and is much wider than the rest of the intestine. The ileum runs posteriorly in a straight line for some distance, after which it curves anteriorly. It once again bends back and then follows a more or less straight course posteriorly, where it dilates into a thick-walled rectum. The length of the gut is about 1.22mms.

XI. *Glyptothorax cavia* (Hamilton)

Glyptothorax cavia is a rheophilous bottom—dwelling teleost which feeds on the benthonic invertebrates such as annelids, the ephemerid, trichopterid and dragon-fly larvae, *Gammarus, Eubranchipus, Triops*, aquatic beetles, etc., together with a small amount of filamentous blue-green algae. The plant food is comprised by *Spirogyra, Oedogonium, Ulothrix*, etc. The animal food-plant food ratio in the average annual feed of the sisorid was found to be 70 : 5, the rest being made up of decaying organic matter (10 per cent) and about 15 per cent of sand and allied inorganic detritus, indicating a benthophagous habit.

Seasonal fluctuations in the average annual feed of the fish have also been noted and are correlated with the relative abundance of various food items in the environment. During the months of April and May it feeds almost exclusively on ephemerid larvae. The incidence of stenophagy during the said months is due to the availability of the ephemerid larvae in the environment of the fish. Similar observations of the incidence of stenophagy were reported by the present author in case of *Glyptothorax kashmirensis* (Nath 1976).

Being a fast-stream dweller, *G. cavia* has reduced eyes covered over by skin as an adaptation to life in turbulent waters, with dashing sand and silt. This sisorid, therefore, depends upon its four pairs of barbels for the location of its food items. Experiments performed on the fish have shown that it is a nocturnal feeder, its sense organs being, therefore, adapted to a nocturnal mode of life. It is a touch-and taste feeder and shares this character with several other torrential-stream sisorid fishes such as *G. kashmirensis, G. conirostre* and *Glyptosternum recticulatum* (Das and Nath 1967).

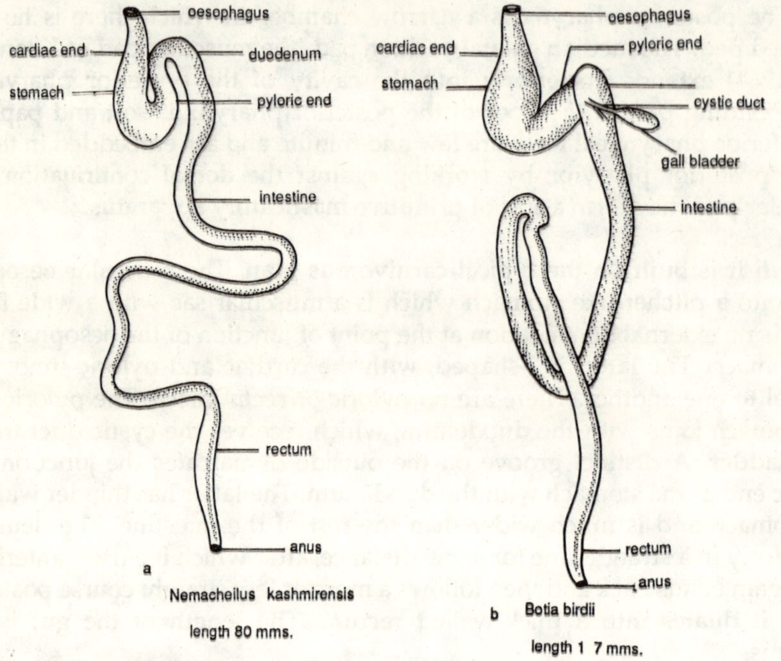

Plate XIV : a. Gut of *Noemacheilus kashmirensis* uncoiled; b. Gut of *Botia birdi* uncoiled

The Oropharynx : The mouth is inferior, transverse, narrow and semi-circular in adaptation to the benthophagous habit. The rostral skin is consequently carried over to the ventral surface to cover the upper (anterior) lip in the form of a rostral hood. The lower (posterior) lip is comparatively narrow and both the lips are thick and fleshy and beset with papillae on their ventral surface for assisting the fish in adhering to the substratum. The lips are supported by well developed jaws, the upper jaw being slightly the longer of the two. Four pairs of barbels (one pair each of maxillary, nasal and two pairs of mandibulars) are present in this sisorid and help in the location of the food items. The maxillary barbels are the longest, with broad bases and reaching slightly beyond the base of the pectoral fins. The outer mandibular barbels reach the gill openings, while the inner mandibulars are equal to the nasal barbels, the nasals equalling the distance between their bases. The large maxillary barbels form an arch along with the rostrum and upper lip and constitute an adhesive-cum-feeding mechanism. The fish picks up food from the substratum by attaching itself closely to the bottom, with an anterior extension of the fleshy upper lip.

The mouth opens into a depressed oropharyngeal cavity, which is narrow anteriorly but widens considerably towards the posterior region. The roof of the oropharyngeal cavity is slightly concave while the floor is nearly flat. The tongue is absent. A well developed, arched maxillary valve of uniform thickness stretches across the upper jaw. The mandibular valve is thin and shorter than the maxillary valve and lies immediately behind the patch of mandibular teeth.

Several sets of villiform teeth occur in the anterior and posterior regions of the oropharynx. In a specimen 110 mm long, the following disposition of teeth was found:

I. Teeth on the roof of the oropharynx:
1. Premaxillary teeth
2. Maxilary teeth
3. Superior-pharyngeal teeth.

On the upper jaw teeth are present on the premaxillae and maxillae in a single band which is crescent-like and measures 8 mm in length and 2 mm in (maximum) width and bears 6-7 rows of teeth. These teeth are villiform, with their edges directed backwards. The maximum size attained by these teeth does not exceed 0.25 mm. The palate is edentulous. Hora (1923) remarked that the broad premaxillary tooth band may be considered as a specific character, separating *G. cavia* from the remaining species under the genus *Glyptothorax* Blyth. The superior pharyngeal teeth are borne on two oval patches, each patch measuring about 5 mm in length and 3 mm in width.

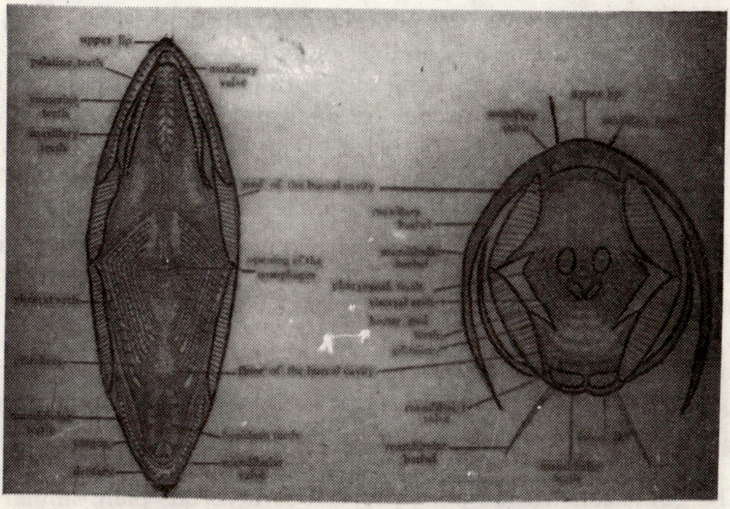

Plate XV : a. Buccopharynx of *Salmol trutta fario* spread wide open
b. Buccopharynx of *Glyptothorax cavia* spread wide open

II. Teeth on the floor of the oropharynx:
1. Mandibular teeth
2. Inferior-pharyngeal teeth.

The villiform mandibular teeth comprise two patches, one on each side of the mandibular symphysis, a feature also met with in *Glyptosternum reticulatum.*

They are arranged in 5-6 rows and are similar to the premaxillary teeth in form and structure. Each patch of mandibular teeth is about 5 mm long and 2 mm broad, with a thin partition between the two patches. A pair of closely-approximated and nearly triangular horny pads occur on the floor of the oropharyngeal cavity and bear small, homodont teeth, the inferior-pharyngeal teeth, which are backwardly directed and work against the superior-pharyngeal teeth present on the roof of the oropharynx. Each patch is about 4 mm long and 2 mm (maximum) in width.

As discussed elsewhere (Nath 1976), the recovery of totally undamaged ephemerid larvae from the stomach of *G. kashmirensis* as well as *G. cavia* indicates that, contrary to the popular belief, the backwardly-directed superior and inferior-pharyngeal teeth are not of much help in the mastication of the ingested food but assist in gripping the live prey and guiding its entry into the oesophagus, and preventing regurgitation. As the tongueless *Glyptothorax* evaginates its cardiac stomach to swallow the large prey as entire, there is no mastication of the food in the oropharyngeal cavity.

The lateral walls of the anterior pharynx bear four pairs of oblique gill slits supported by the visceral arches, which are provided with gill rakers. The gill rakers are long, pointed, fairly stiff and inwardly directed. The gill rakers occur only on the anterior face of the first two pairs of branchial arches; the third and fourth pairs of branchial arches bear gill rakers on both sides, while the fifth pair of branchial arches bear them only on their outer face. The gill rakers of the first two pairs of branchial arches are long and stiff and assist in holding the live prey, while those of the last two arches are small and knob-like and interdigitate to prevent the escape of the food particles along with outgoing water current. The maximum length of the gill rakers was about 3 mm in a specimen 110 mm long.

The posterior pharynx is a narrow triangular chamber which does not contain any horny pad on its roof, as is met with in several herbivorous as well as omnivorous teleosts such as *Labeo dero, Crossocheilus latius punjabensis, Cyprinus carpio*, etc. There is little mastication of food in the posterior pharynx and the villiform superior- and inferior-pharyngeal teeth borne on the horny pads assist in gripping the food, which is living and actively struggling, on its way to the oesophagus.

The Gut: The gut is constructed on the typical carnivorous plan, similar to that described in case of *G. kashmirensis* and *Glyptosternum reticulatum* (Das and Nath 1967, 1973). The muscular oesophagus leads into a well developed, thick-walled pitcher-like stomach, which is roughly J-shaped, with a well developed fundus. There is no external demarcation between the oesophagus and the cardiac end of the stomach but there is a well marked constriction at the pyloric end of the stomach at the point of its junction with the duodenum. There are no pyloric caeca. The duodenum is a wide, tubular structure which overlies the stomach ventrally and has relatively thinner walls than the stomach. The intestine is thrown into a few convolutions and then dilates into the rectum. The relative length of the gut is about 1.12mms.

DISCUSSION

The feeding habits of the adult fish become apparent when a single organ or a few organs are examined or, sometimes, only when every organ of the digestive system has been examined (Suyehiro 1942). Many studies have been made on the extent of adaptation of the alimentary canal in fully grown teleosts to a particular kind of natural diet. Ghazzawi (1935), Suyehiro (1942), Al-Hussaini (1945, 1946, 1947, 1949), Angelescu and Gneri (1949), Girgis (1952), Kapoor (1953, 1957, 1958), Pillay (1953), Khanna (1961), Mohsin (1962), Lal *et al.* (1964), Sehgal (1966), Lal (1968), Schmitz and Baker (1969), Sehgal and Salaria (1970), Sriwastava (1970), Bucke (1971), and Verma *et al.* (1974) have found that different species with the same type of diet may differ in the structure of the alimentary canal, but the functional adaptations related to the nature of food and feeding habits remain similar, although the degree of relation between digestive tract and food varies. Harder (1960) and Bishop and Odense (1966) have, however, raised doubt as to the relation between feeding habits and gut morphology.

It is known that the structure of the digestive apparatus is related to the form of the body. Besides the food and feeding habits, the phylogeny of a species is another important factor in the final construction of the digestive apparatus.

It is a well established fact that the herbivorous fishes have a longer alimentary canal than that of the carnivorous species. This is true not only for fishes but also for many other groups of vertebrates. The relative length of the gut shows a well-marked gradation from the herbivorous to the carnivorous condition, as has been found by Das and Moitra (1955, 1956, 1958, 1963) in 24 species of freshwater teleosts from Uttar Pradesh and also by the present author in 19 species of teleosts from Jammu and Kashmir.

In the carnivorous fishes the length of the gut is almost equal to or, in some cases, less than the body length, while in the herbivores the length of the gut is much longer than the body length. In omnivores the gut length varies between that of herbivores and carnivores (Tabel 1).

Table 1 : Feeding Habits and Relative Length of the Gut in Some Fishes of Jammu Province

Sl.No.	*Species*	*Feeding Habits*	*Animal food-Plant food Ratio*	*R.L.G.*
1.	*Crossocheilus latius punjabensis* Mukerji	Herbivore	11 : 85	9.14
2.	*Labeo dero* (Ham)	Herbivore	15 : 80	8.86
3.	*Garra gotyla gotyla* (Gray)	Herbivore	5 : 85	8.19
4.	*C. carpio specularis* Lacep.	Herbivore	20 : 65	2.00
5.	*Cyprinus carpio communis* Lin.	Herbivore	34 : 57	1.92
6.	*Rasbora (Rasbora) daniconius daniconius* (Ham.)	Omnivore	65 : 35	1.10
7.	*Barilius vagra vagra* (Ham.)	Omnivore	70 : 30	0.56
8.	*Puntius conchonisu* (Ham.)	Omnivore	30 : 50	3.25
9.	*Noemacheilus rupecola rupecola* (McCl.)	Carnivore	70 : 20	0.88
10.	*N. vittatus* (Heckel)	Carnivore	85 : 10	1.22
11.	*Glyptothorax cavia* (Ham.)	Carnivore	70 : 5	1.12

The adaptations of the alimentary canal of fishes to their food are particularly evident in the form of mouth, size, shape and structure of the oropharynx, the dentition, the gill rakers, the structure of the oesophagus, the presence or absence and shape of the stomach and the intestine and the relative length of the gut. All these structures are subject to diverse and significant variations and modifications in accordance with the feeding habits of different species.

The diversity in feeding habits that the fishes exhibit is partially the result of evolution leading to the structural adaptations for getting food from the equally great diversity of situations that have evolved in the environment. Conversely, the importance of food in the daily life of a fish is "reflected" in the form of mouth and jaws, dentition, the shape and size of gill rakers, etc., and, therefore, the differences exhibited by these structures are actually the morphological expressions of the differences in their feeding habits.

Suyehiro (1942) concluded that the digestive organs of fishes depend upon their phylogeny as well as the feeding habits and consequently fishes with identical feeding habits may not necessarily possess similar digestive organs.

Crossocheilus latius punjabensis, Labeo dero, Garra gotyla gotyla, Cyprinus carpio specularis and *C. carpio communis* are herbivorous teleosts. In these fishes the mouth is bounded by soft lips supported by edentulous jaws. As the food mainly consists of soft, aquatic vegetation such as blue-green filamentous algae and encrusting periphyton (*Garra*), there is no need for the presence of a cutting or crushing apparatus. Therefore, the anterior pharynx is entirely devoid of teeth. The anterior-pharyngeal teeth also are absent in these herbivorous fishes. In *Crossocheilus* the upper lip is fringed and the lower lip studded with fleshy papillae on its ventral surface to serve as an adhesive organ.

Rasbora daniconius, Barilius vagra vagra and *Puntius conchonius* are omnivorous species. In these fishes the mouth is bounded by soft, unfringed lips. The jaws are edentulous and teeth are also absent from the oropharynx. The mouth is oblique and moderately wide in *Rasbora* and the lower jaw is slightly protruding; the rather uppishly directed mouth indicates the capacity of the fish to obtain food directly from the surface. In *Barilius* the mouth is somewhat oblique and the lower jaw usually has a knob at the symphysis, the upper jaw bearing a corresponding emargination to receive the knob, a characteristic feature of predatory carps, such as *Aspius aspius, Barilius moore* and *B. christy* (Kazansky 1964, Matthes 1963).

Noemacheilus rupecola rupecola, N. vittatus and *Glyptothorax cavia* are carnivorous fishes. *Noemacheilus* possesses a subterminal mouth which is bordered by thick, papillated lips, this being an adaptation to life in the fast streams, as the lips, when pulled outwards and away from the mouth, provide a ring-like adhesive sucker to enable the loach to affix itself firmly to some submerged object and stay out of the way of mountain torrents. The jaws are edentulous. In *Glyptothorax* villiform teeth are present on the maxillaries and the dentaries; besides these, teeth are also present on the roof of the posterior-pharynx (superior-pharyngeal teeth) and on the horny pads present ventrally; however, these teeth do not help in mastication of the food but assist in gripping the live prey and guiding its entry into the oesophagus. The mouth is inferior, transverse and

Table 2 : Percentage Composition of the Food of Fishes

	Species	*Animal Food*								*Plant Food*				*Miscellaneous and DOM*			*RATIO*			
		Protozoa	*Rotifers*	*Copepoda*	*Cladocera*	*Ostracoda*	*Insect larvae*	*Insects*	*Annelids, Molluscs etc.*	*Green algae*	*Diatoms*	*Macrophytes*	*Unidentified*	*Fish scales*	*Fish eggs.*	*Sand & mud*	*Animal*	*Plant*	*Miscellaneous*	*Total*
1.	*Crossocheilus latius punjabensis* Mukerji	1	1	2	2	1	1	1	2	40	7	33	5	.5	1	2.5	11	85	4	100
2.	*Labeo dero* (Hamilton)	5	1	2	2	1	2	1	1	46	10	22	2	.5	.5	4.0	15	80	5	100
3.	*Garra gotyla gotyla* (Gray)	1	1	.5	.5	1	.5	.5	—	50	10	20	5	1	.5	8.5	5	85	10	100
4.	*Cyprinus carpio specularis* Lacepede	3	4	4	2	1	1	1	4	20	8	35	2	3	4	8	20	65	15	100
5.	*Cyprinus carpio communis* Linnaeus	3	5	8	5	2	3	5	3	15	7	34	1	2	1	6	34	57	9	100
6.	*Rasbora daniconius daniconius* (Hamilton)	5	5	10	10	5	12	15	3	10	8	15	2	—	—	—	65	35	—	100
7.	*Barilius vagra vagra* (Ham.)	5	10	15	10	5	10	10	5	10	8	10	2	—	—	—	70	30	—	100
8.	*Puntius conchonius* (Ham.)	5	7	2	7.6	3	3	2.5	4.4	15	15	23	5	5	3	4	30	58	12	100
9.	*Noemacheilus rupecola rupecola* (McCland).	3	6	15	10	15	10	7	4	10	5	4	1	1	2	7	70	20	10	100
10.	*Noemacheilus vittatus* (Heckel)	2	3	10	16	16	15	20	3	7	2	1	1	1	2	2	85	10	5	100
11.	*Glyptothorax cavia* (Hamilton).	1	2	10	12	5	20	15	5	2	1	1.5	.5	4	—	21	70	5	25	100

Table 3 : Feeding Habits and Morphology of the Mouth Parts and Associated Structures in Some Fishes of Jammu Province

Sl. No.	*Species*	*Feeding habits*	*Mouth and lips*	*Dentition*	*Palatal organs*	*Gill rakers*
1.	*Crossocheilus latius punjabensis* (Mukerji)	Benthophagous Herbivore.	Mouth sub-terminal; lips soft, fleshy, upper fringed, lower studded with fleshy papillae ventrally	Jaws and palate edentulous; small, crooked inferior-pharyngeal teeth work against a dorsal cornified pad for crushing food material.	A dorsal lamellar organ bearing fleshy papillae present; a well developed palatal pad present behind the lamellar organ and studded with papillae for respiration.	Gill rakers, Short, fine, soft and hair-like, closely set on both sides of branchial arches, interlocking for sieving of small food particles.
2.	*Labeo dero* (Hamilton)	Benthophagous Herbivore; resorts to plankton & mid-feeding during day.	Mouth narrow, transverse, sub-terminal; lips fleshy, fringed, non-protractile, lined internally by a cartilaginous base.	Jaws and palate edentulous; strong, plough-shaped inferior-pharyngeal teeth work against a well developed dorsal cornified pad for crushing molluscs and insects.	A dorsal lameller organ as well as a palatal organ present, studded with papillae.	Gill rakers short, fine, almost hair like and closely set on both sides of branchial arches inter-digitate in the contracted condition to form a sieve.
3.	*Garra gotyla gotyla* (Gray).	Benthophagous herbivore.	Mouth inferior, transverse and semicircular; lips thick, broad, upper fringed, lower arched and lined internally by a cartilaginous cutting edge for scrapping entrusting periphyton.	Jaws and palate edentulous; well developed uncinate inferior pharyngeal teeth work against the dorsal cornified pad for mastication of food.	Only a palatal organ present, studded with respiratory papillae; Lamellar organ absent.	Gill rakers few, slender, short, hair like and closely set.

4.	*Cyprinus carpio specularis* Lacepede	Planktophagous Herbivore; also seeks food from the bottom.	Mouth Protrusible, terminal, oblique and depressed; lips soft and infringed.	Jaws and palate edentulous; several prominent molariform inferior pharyngeal teeth work against a well developed dorsal cornifred pad for crushing mollusce shells.	Lamellar organ absent. Palatal cushion pad thick and muscular, studded with knob-like papillae.	Gill rakers fairly long and hair like, closely set on both sides of the branchial arches; gill rakers of adjacent arches interdigitate for sieving out plankton.
5.	*Cyprinus carpio communis* Linnaeus.	Planktophagous herbivore; also sucks food from the bottom.	Mouth terminal and oblique; lips soft, fleshy and unfringed.	Jaws and palate edentulous; several strong molariform inferior-pharyngeal teeth and well developed cornified pad present.	Lamellar organ absent; palatal cushion pad fleshy and well developed and studded with knob-like papillae.	Gill rakers fairly long and soft, closely set on both sides of the gill arches and interdigitating for sieving plankton.
6.	*Rasbora daniconius daniconius* (Hamilton).	Carni-Omnivore.	Mouth oblique, moderately wide; lower jaw slightly protruding; lips thin, soft and unfringed.	Jaws and palate edentulous; small, sharp and curved inferior-pharyngeal teeth present; cornified pad poorly developed.	Lamellar organ absent; palatal cushion pad well developed and studded with large respiratory papillae.	Gill rakers short, lanceolate, closely set on both sides of the gill arches and interdigitating for sieving micro-organisms.
7.	*Barilius vagra vagra* (Hamilton)	Predatory Carni-Omnivore	Mouth non-protractile, anterior, somewhat oblique, with a deep cleft; lips soft, thin and unfringed.	Jaws and palate edentulous; crooked, minute inferior-pharyngeal teeth and reduced cornified pad present.	Lamellar organ absent; palatal cushion pad poorly developed and studded with gustatory papillae.	Gill rakers very short, soft and closely set on both sides of gill arches; rakers of adjacent arches nearly interdigitating.

contd. . . .

S.No.	Species	Feeding habits	Mouth and lips	Dentition	Palatal organs	Gill rakers
8.	*Puntius conchonius* (Hamilton)	Herbiomnivore	Mouth arched, anterior, and non-protrusible; lips soft, thin, unfringed and crescentic, without a horny covering.	Jaws and palate edentulous; small inferior pharyngeal teeth and dorsal cornified pad present.	Lamellar organ and cushion pad absent.	Gill rakers, produced into thorn-like projections and interdigitating for sieving.
9.	*Noemacheilus rupecola rupecola* (McClelland)	Entomophagous Benthophage.	Mouth small, inferior, transverse and subterminal, lips thick, papillated, lower lip interrupted in the middle.	Jaws and palate edentulous; inferior-pharyngeal teeth minute and few. No cornified pad.	Lameller organ absent; a muscular pad present on palate.	Gill rakers soft, small, stumpy, closely set and interdigitating for sieving out minute food particles.
10.	*N. vittatus* (Heckel)	Benthophagous entomophage.	Mouth small, inferior, transverse, and subterminal; lips thick and papilalted.	Jaws and palate, edentulous; inferior-pharyngeal teeth minute and few. No cornified pad.	Lamellar organ absent; a reduced muscular pad present on palate.	Gill rakers slender and spine like.
11.	*Glyptothorax cavia* (Hamilton)	Benthophagous carnivore.	Mouth inferior, transverse, narrow and semicircular; lips thick, fleshy and beset with papillae on their ventral surface for adhesion; lower lip narrower.	Jaws bear villiform teeth, borne in single crescentic band on the upper jaw and in two separate patches on the mandibles; superior-and inferior-pharyngeal teeth present and borne on two pairs of horny pads.	Lamellar organ and palatal cushion pad absent.	Gill rakers long, pointed, fairly stiff and inwardly directed for assisting in holding the live prey; gill rakers of last two arches interdigitate for preventing escape of food particles.

semicircular, an adaptation to the benthophagous habit. The lips are beset with papillae on their ventral surface for assisting the fish in adhering to the substratum. In *Noemachilus* the inferior-pharyngeal teeth alone are present and are few and minute.

In *Garra, Cyprinus, Rasbora, Barilius, Puntius, Noemacheilus* and *Glyptothorax* the anterior-palatal pad (dorsal lamellar organ) is totally absent. This pad is present in the herbivorous teleosts *Crossocheilus* and *Labeo*. It is significant to note that such an organ is absent in *Garra* but present in *Crossocheilus*, although both the fishes belong to the same subfamily (Garrinae). Majumdar and Saxena (1961) studied the palatal organs of *Labeo dero* and concluded that carps with modified palatal surface can survive for a considerably longer time outside water than those without such palatal structures. Moreover, bottom dwellers inhabit an environment which is always deficient in dissolved oxygen; hence bottom-dwelling carps require a greater area to make up for the deficiency of oxygen, which is provided by the palatal pads and their papillae in addition to the normal gills. The skin, being covered over with scales, cannot presumably function as a respiratory organ. Longer duration of life of these carps even outside water is made possible by the oxygen absorbed through these palatal structures and partly through the gills. The results of experiments performed on *Crossocheilus* and also the vascularity of the palatal organs and their papillae point to their respiratory function.

A posterior, median-palatal, cushion-pad is present in the herbivorous, omnivorous as well as the insectivorous fish species studied, being found in *Crossocheilus, Labeo, Garra, Cyprinus, Rasbora, Barilius, Puntius* and *Noemacheilus*, but is absent in *Glyptothorax*. It may, however be pointed out here that the posterior-palatal (cushion) pad does not attain the same high degree of development and structural elaboration in the carnivorous forms (*Noemacheilus*) that it does attain in the herbivorous and omnivorous species discussed in the present contribution.

Popta (1900) found that the gill rakers are specifically peculiar and diagnostic in the various species of teleosts examined by him. He interpreted the arrangement of gill rakers and their development in relation to the form of mouth and the nature of food. Zander (1903) examined the gill rakers of a large number of fishes and observed that in the carnivores (e.g. *Esox, Lucioperca*) the gill rakers are in the form of teeth. In many other genera the gill rakers form a sieve like filtering apparatus (*siebfortsatze*) which is developed in some cases upon both edges of the basibranchial arches, or, in others, only upon the anterior edges. He also described the relative fitness of the "*siebfortsatze*", which is correlated with the nature of food of the species where they occur and attains extreme development in this respect among the planktophagous fishes.

Das and Moitra (1955, 1956, 1958), in their studies on the alimentary canal of 24 species of teleosts from Northern India, observed that the gill rakers are variously modified according to the feeding habits of the fish. "In herbivorous species the gill rakers are thin and long and partially adapted to perform a sieving function. In the omnivorous species the gill rakers are in the form of teeth-like processes developed into protective elements to prevent the blocking

Table 4 : Feeding and Gut Correlations

Sl.No.	Species	Feeding habits	Oesophagus	Stomach	Intestine	Pyloriccaeca	Rectalcaeca
1.	*Crossocheilus latius punjabensis* Mukerjii	Herbivore	Short and muscular	x	Long and coiled, with well developed intestinal bulb.	x	x
2.	*Labeo dero* (Hamilton)	Herbivore	Short and thick walled	x	Long and coiled; intestinal bulb well developed.	x	x
3.	*Garra gotyla gotyla* (Gray)	Herbivore	Short, tubular and muscular	x	Long and coiled like a watchspring; intestinal bulb well developed.	x	x
4.	*Cyprinus carpio specularis* Lacepede	Herbivore	Short and muscular	x	Long and thin walled, with few coils; intestinal bulb well developed.	x	x
5.	*C. carpio communis* Linnaeus	Herbivore	Short and muscular	x	Long and thrown into a few coils; duodenum hypertrophied to form well developed intestinalbulb.	x	x
6.	*Rasbora daniconius daniconius* (Hamilton.)	Omnivore	Short and muscular	x	Short with well developed, intestinal bulb; very few coils	x	x
7.	*Barilius vagra vagra* (Hamilton)	Omnivore	Short and muscular	x	Short, intestinal bulb well developed.	x	x
8.	*Puntius conchonius* (Hamilton)	Omnivore	Short and muscular	x	Long and coiled; intestinal bulb well developed.	x	x

9.	*Noemecheilus rupecola rupecola* (McClelland)	Carnivore	Short and muscular	Muscular and pitcher like, with wide fundus and well differentiated cardiach and pyloric regions.	Short, with very few coils; duodenum externally demarcated from the pyloric limb of the stomach by a conspicuous constriction.	x	x
10.	*N. vittatus* (Heckel)	Carnivore	Short and muscular	Muscular and pitcher like with wide fundus and well developed cardiac and pyloric regions.	Short, with a few coils; duodenum and pyloric limb of the stomach externally demarcated by a well marked constriction.	x	x
11.	*Glyptothorax cavia* (Hamilton)	Carnivore	Narrow and muscular	Muscular and well developed; not differentiated externally into cardiac and pyloric regions.	Short and thick walled; duodenal portion well marked off from the rest of the intestine.	x	x

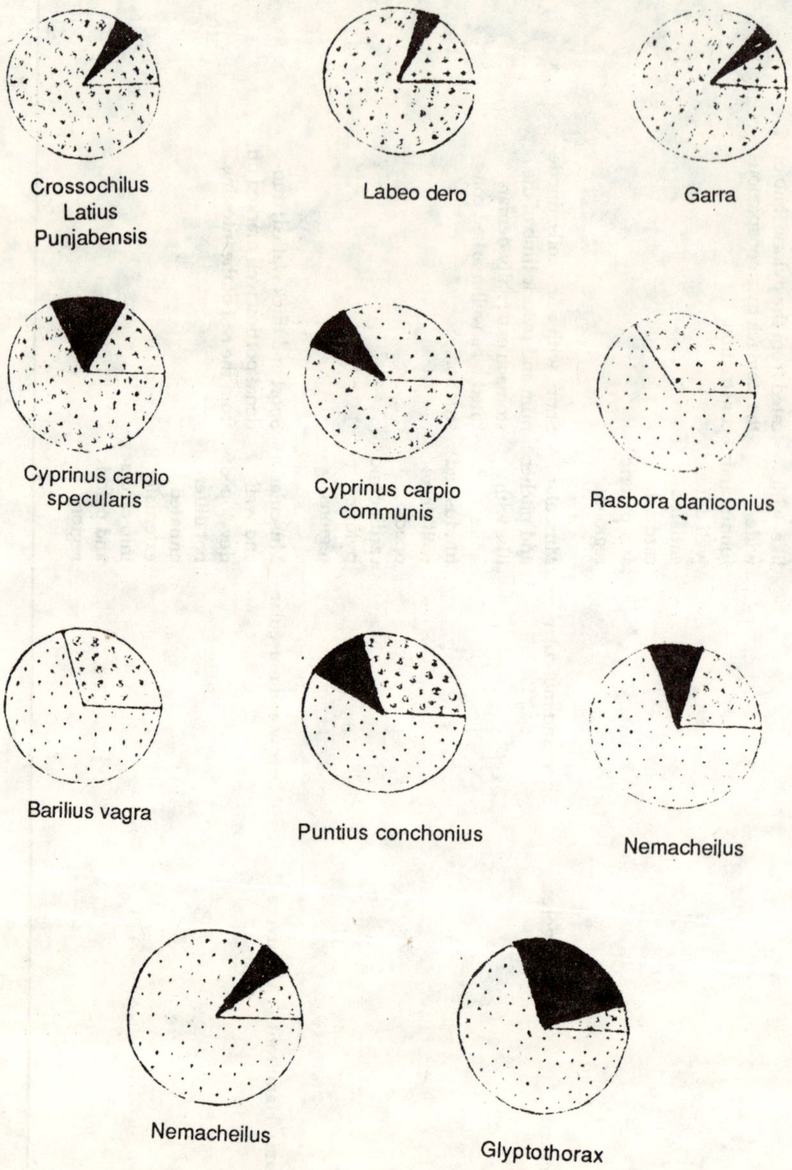

Plate XVI : Food spectrum of some fishes of Jammu Province

A = Detritus B = Plant food X X X X X C = Animal food

of the gillslits and also to provide additional hold on the living, struggling prey. The original, sieving function of the gill rakers is thus entirely lost with the change to even part-carnivore diet. In the carnivorous species the gill rakers are modified into flat and round plate-like structures provided with short, conical teeth".

Kozhina (1958) observed that the gill rakers of mature predatory fish are coarse, sparse, short and armed with small, sharp denticles. Fish feeding on plankton and microbenthos have long, thin, soft and densely arranged gill rakers, forming a net for filtering out small organisms. The benthophages have rod-shaped gill rakers drawn out into cones. This corresponds to their feeding, because, together with benthonic organisms they swallow the soil in which the organisms live. Fish feeding to a considerable extent on vegetable food possess short, sparse, rod-shaped gill rakers and a thick gill arch.

In the present studies on the eleven species of freshwater teleosts from Jammu Province, considerably varied modifications of the gill rakers have been observed which can be definitely attributed to the differences in the feeding habits of these fishes. In the herbivorous species, viz., *Crossocheilus, Labeo* and *Garra*, the gill rakers are short and thin and closely set on both sides of the visceral arches while in *Cyprinus* they are long and soft. The gill rakers of the adjacent arches interdigitate for sieving out the mud and sand as in the planktophagous fishes. The long gill rakers are characteristic for the majority of bottom feeders which are in the habit of stirring up mud at the bottom (Kyle 1926).

In the omnivorous species, viz., *Rasbora, Barilius* and *Puntius*, the gill rakers are short, fine, almost hair-like and occur on both sides of the branchial arches. They are closely set and interdigitate with those of the adjacent arches for sieving out plankton.

In the carnivorous fishes, viz., *Noemacheilus* and *Glyptothorax*, the gill rakers are considerably modified and adapted to prevent the escape of the prey, as well as for rasping the food. In *Noemacheilus rupecola* the gill rakers are soft, small and stumpy structures, closely set on both sides of the second, third and fourth pairs of branchial arches. The first pair of branchial arches bears gill rakers on one side only. In *N. vittatus* the gill rakers are slender and spine-like and relatively fewer in number. In *Glyptothorax* the gill rakers are long, pointed and fairly stiff and occur on only the anterior face of each branchial arch. They are fully capable of holding the prey and preventing its escape through the gill slits.

A posterior-pharyngeal masticatory apparatus is present both in the herbivorous as well as the omnivorous teleosts studied by the present author, being found in *Crossocheilus, Labeo, Garra, Cyprinus, Rasbora, Barilius* and *Puntius* and compensates for the absence of a true stomach in these teleosts. The herbivorous species (viz., *Garra, Labeo* and *Cyprinus*) possess well-developed inferior-pharyngeal teeth, firmly affixed to the modified fifth branchial arch, which oppose a dorsal cornified pad borne by the special (masticatory) process of the basioccipital. In *Crossocheilus* the inferior-pharyngeal teeth are minute. The mastication of the ingested food is thus effected in the posterior-pharyngeal region, without interfering with respiration in the anterior pharynx. The omnivores—*Rasbora, Barilius* and *Puntius*—possess a relatively degenerate dorsal cornified pad against which work the smaller inferior-pharyngeal teeth. In the insectivorous *Noemacheilus* the dorsal cornified pad is absent and a primitive type of masticatory apparatus is formed by the posterior continuation of the muscular pad, against which work the few and minute inferior-pharyngeal teeth. However, all such structures are totally absent in the truly carnivorous *Glyptothorax*, in which very little mastication occurs in the posterior pharynx and the live prey is swallowed whole.

In the herbivorous teleosts the length of the gut is always much more than the total body length. This is due to the presence of an excess of cellulose matter in their diet, which requires a relatively longer intestine for proper assimilation (e.g. *Crossocheilus, Labeo, Garra* and *Cyprinus*). In the carnivores, the length of the gut is almost equal to or, in some cases, less than the body length, as the animal diet consumed by them requires a relatively much smaller absorptive surface (e.g. *Noemacheilus* and *Glyptothorax*). In the omnivores the gut length varies between that of the herbivores and the carnivores (e.g. *Rasbora, Barilius* and *Puntius*).

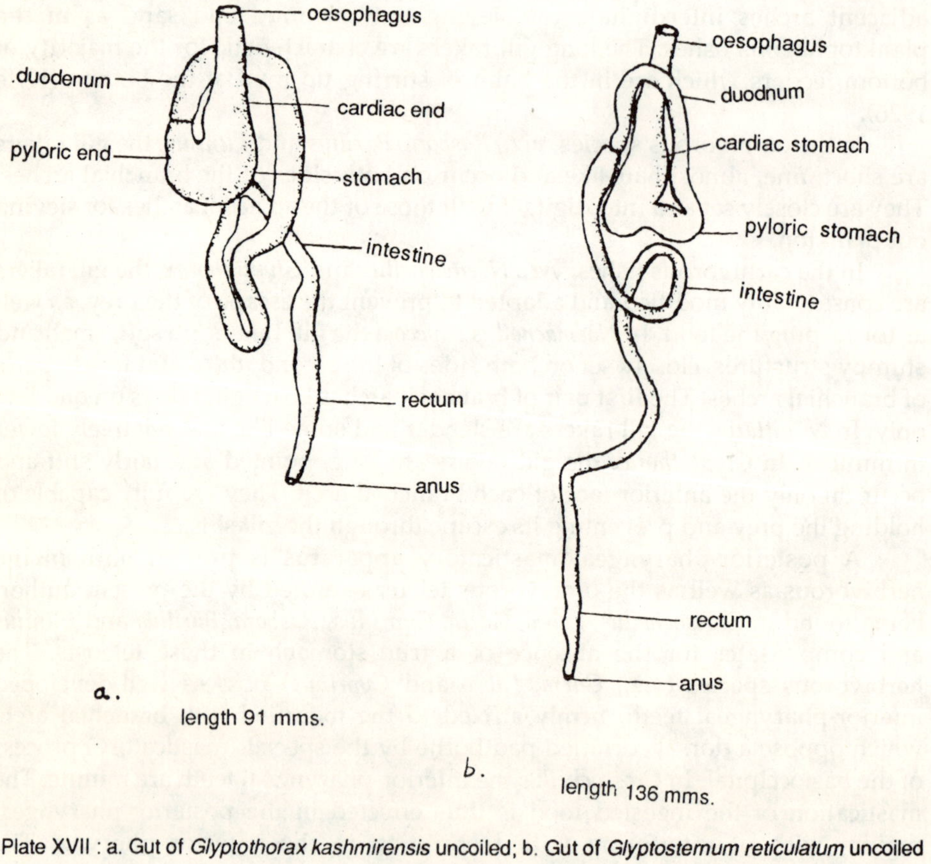

Plate XVII : a. Gut of *Glyptothorax kashmirensis* uncoiled; b. Gut of *Glyptosternum reticulatum* uncoiled

In the herbivorous the intestine is long and highly convoluted and the first limb of the intestine (i.e. duodenum) is hypertrophied to form a dilated bag-like structure—the intestinal bulb. In the absence of the true stomach, the intestinal bulb acts as a storage organ for the huge quantities of aquatic vegetation ingested by the fish (e.g. *Crossocheilus, Labeo, Garra* and *Cyprinus*).

In the omnivorous teleosts (*Puntius, Barilius* and *Rasbora*) the gut shows a gradation from the herbivorous to the carnivorous condition, as the gut length is considerably less than in the case of the herbivores. However, here also a well developed intestinal bulb is present which serves as a storage organ for the ingested food in the absence of a true stomach.

In the carnivorous species (e.g. *Noemacheilus* and *Glyptothorax*) there is a marked further decrease in the relative length of the gut. Other modifications which can be attributed to the almost complete change in feeding habits from the herbivorous to the carnivorous state are the high elasticity of the wall of the oesophagus and stomach, and the presence of a well developed stomach (with strong acidic secretions).

A well developed muscular sac-like stomach is present in all the carnivorous teleosts, and has well-differentiated cardiac and pyloric limbs with a fairly wide fundus. The stomach is totally absent in the herbivores and omnivores included in the present study, being functionally substituted by the intestinal bulb, which represents the hypertrophied duodenum and is not homologous with a true stomach in any way.

Several hypotheses have been put forth to explain the absence of the stomach in teleosts. Karpevitch (1936) opined that the cause of the disappearance of the stomach in some marine fishes is the alkalification of the gastric juice by the sea water swallowed with the food. This alkalification inhibits the peptic activity and leads to the loss of gastric digestion. Jacobshagen (1937) came to the conclusion that the explanation of the absence of a stomach cannot be based on the kind of food ingested. Hirsch (1950) held that the presence of a masticatory apparatus makes that a superfluity and consequently the stomach disappears.

Szarski (1956) suggested that the absence of a stomach in various groups of fish is probably a result of adaptation to the ingestion of food organism with calcareous shells. The alkalification tends to neutralise the acid contents of the stomach and confines digestion to an alkaline medium which might eventually lead to reduction of the stomach. According to Lagler *et al.* (1962) the primary criterion for being able to do without a stomach does not seem to be whether a fish is a herbivore, an omnivore or a carnivore but whether accessory adaptations for trituration and very fine grinding of the food exist, either in the form of teeth or a grinding apparatus (e.g. gizzard). It may be stated here that a perfect posterior-pharyngeal grinding mechanism is present in *Cyprinus, Labeo, Garra* and *Crossocheilus*, etc. and consists of a dorsal cornified pad borne on the masticatory process of the basioccipital, against which work the strongly developed inferior-pharyngeal teeth present on the pharyngeal bone (the *os-pharyngeus* of Goodrich). In *Rasbora, Barilius* and *Puntius* the posterior-pharyngeal masticatory mechanism is rather weak, as the dorsal cornified pad and the inferior-pharyngeal teeth are poorly developed. The foodstuffs ingested by these teleosts do not, however, require much trituration before passing into the intestinal bulb.

Verighina (1977) is of the opinion that the nature of food consumption should be considered as one of the factors that, in the course of evolution, led to a gradual reduction in the size of the stomach and its eventual disappearance in the herbivorous teleosts.

In all the eleven species of teleosts studied in the present work the pyloric caeca are totally absent. It has been established by Al-Hussaini (1947) that the presence and absence of the pyloric caeca cannot be taken as a significant factor in relation to the feeding habits of a fish, as these structures have been found to be present both in the carnivores as well as the omnivores. Khanna (1961) suggested that the function of the pyloric caeca is simply to increase the absorptive surface of the alimentary canal for the sake of proper assimilation of the digested food. It may, however, be pointed out here that the pyloric ceca have been found to be present only in omnivorous and carnivorous teleosts (Das and Moitra 1956; Das and Nath 1965, 1967) and the presence of digestive enzymes in the caeca has also been demonstrated.

The duodenum and ileum are not externally demarcated in *Noemacheilus* and *Glyptothorax*, all of which are carnivorous teleosts. The rectum is not clearly defined externally in any of the species studied by the present author, although internally the rectal part of the gut can be recognised by the differences in the histology and the structure and disposition of the mucosal folds. The rectal caeca also are absent in the teleosts under discussion.

REFERENCES

Agrawal, V.P. and Bala, B. 1967. Studies on the morphology and physiology of the alimentary canal of *Mugil corsula* (Ham.) *Agra Univ. J. Res. (Sci.)*, **16** : 107-120.

Agrawal, V.P. and Sharma, Usha 1966. Morpho-histological studies of the digestive tract of *Mystus vittatus* (Bloch). *Proc. natn. Acad. Sci. India*, **36** B (4) : 441-456.

Agrawal, V.P. and Tyagi, M.P. 1969. Food, feeding habits and the alimentary canal of freshwater fishes of Muzaffarnagar. *Agra Univ. J. Res.*, **18** : 15-28.

Alexander, R. McN. 1966. The functions and mechanisms of the protrusible upper jaws of two species of cyprinid fishes. *J. Zool. Soc. London*, **149** : 288-296.

Alexander, R. McN. 1967. The functions and mechanisms of the protrusible upper jaws of some Acanthopterygian fish. *Ibid.*, **151** : 43-64.

Alexander, R. McN. 1969. Mechanics of the feeding action in a Cyprinid fish. *Ibid.*, **159** : 1-15.

Alexander, R. McN. 1970. Mechanics of the feeding action of voracious teleost fishes. *Ibid.*, **162** : 145-156.

Al-Hussaini, A.H. 1945. The anatomy and histology of the alimentary tract of the coral feeding fish, *Scarus sordidus* (Klunz.). *Bull. Inst. Egypte*, **27** : 349-377.

Al-Hussaini, A.H. 1946. The anatomy and histology of the alimentary tract of the bottom-feeder, *Mulloides auriflamma* (Frosk.). *J. Morph.*, **78** (1) : 121-153.

Al-Hussaini, A.H. 1947a. The anatomy and histology of the alimentary tract of the plankton feeder, *Atherina forskali* Rupp., *Ibid.*, **80** : 251-286.

Al-Hussaini, A.H. 1947b. The feeding habits and the morphology of the alimentary tract of some teleosts living in the neighbourhood of the Marine Biological Station, Ghardaqa, Red Sea *Publ. Mar. Biol. Sta. Ghardaqa (Red Sea)*, No. **5** : 4-61.

Al-Hussaini, A.H. 1949. On the functional morphology of the alimentary tract of some fish in relation to differences in their feeding habits : Anatomy and histology. *Q. J. Micr. Sci.*, **90** : 109-139.

Al-Hussaini, A.H. 1951. The feeding habits and guts of teleosts, especially of the Northern Red Sea. *Revue. Faculte Sci. Univ. Istanbul.* Ser. *B. Sciences Nat., Fasc.* **2** : 121-129.

Al-Hussaini, A.H. and Kholly, A.A. 1953. On the functional morphology of the alimentary tract of the some omnivorous teleost fish. *Proc. Egypt. Acad. Sci. Cairo*, **9** : 17-39.

Alkholy, A.A. and Abdel Malek, S.A. 1972. Food and feeding habits of some Egyptian fishes in Lake Quarum I. *Tilapia zilli* (Gerv.) A. According to different localities. *Bull. Inst. Oceanogr. Fish. UAR*, **2** : 185-202.

Andriashev, A.P. 1945. On the function of the pharyngeal apparatus in some pelagic predatory fish. *Priroda*, No. **2** : 68.

Andriashev, A.P. 1946. On the simplified structure and function of the pharyngeal apparatus in some herbivorous fish. *Zoologicheskii Zhurnal*, **25** **(4)** : 339-346.

Andriashev, A.P. 1946 b. On the degeneration of structure and function of the pharyngeal complex of some phytophagous fishes. *Ibid*., **25** **(4)** : 345-346.

Anthony, A.D. 1983-84. Food and feeding habits of *Garra mullya* (Sykes) and *G. mcclellandi* (Jerdon). *Matsya*, **9-10** : 151-156.

Aravindan, C.M. 1980. Food selection and feeding habits of *Tilapia mossambica* Peters in different ecological habitats. *Proc. Indian natn. Sci. Acad*., **B 46** (1) : 54-57.

Badola, S.P. and Singh, H .R. 1978. Ecological studies on the hill-stream fishes of the Garhwal Himalaya I. Food and feeding habits of *Schizothorax* sps. In : *Seminar on "Mountain Ecology"*, March 6-8, 1978, Dehra Dun.

Badola, S.P. and Singh, H.R. 1980. Food and feeding habits of fishes of the genera *Tor*, *Puntius* and *Barilius*. *Proc. Indian Nat. Sci. Acad*., **46** (1) : 58-62.

Belonozhko, R.A. 1967. A study of the process of adaptation of the gut to seasonal changes of diet. In : *Fiziologiya i patologiya organov pihevareniya* : 28-30, *Yaroslavl* (In Russian).

Berry, P.Y. and Low, M.P. 1970. Comparative studies on some aspects of the morphology and histology of *Ctenopharyngodon idellus*, *Aristichthys nobilis* and their hybrid (Cyprinidae) *Copeia*, No. **4** : 708-726.

Bertmar, G., Kapoor, B.G. and Miller, R.V. 1969. Epibranchial organs in lower teleostean fishes—an example of structural adaptation. In : *International Review of General and Experimental Zoology*: (W.J.L. Felts and R.J. Harrison, Eds.) Vol. **IV** : 1-48, Academic Press, New York.

Biedermann, W. 1911. *Handbook der Vergleichenden Physiology*. Ed. by H. Winterstein, Bd. **2**, Halfte **I**, Jena.

Bishop, C. and Odense, P.H. 1966. Morphology of the digestive tract of the cod—*Gadus morhua*. *J. Fish, Res. Bd. Canada*, **23** : 1607-1613.

Biswas, S.P. 1986. Food and feeding habits of a minor carp, *Labeo dero* (Ham.). *Proc. 73rd Indian Sci. Congr*., III (7) : 44.

Biswas, S. and Nassar, S.A.K. 1981. Studies on the gut length in relation to feeding habits of five hillstream fishes in Assam. *Proc. 68th. Indian Sci. Congr*., III (7) : 16-17.

Blaber, S.J.M. 1876. The food and feeding ecology of the Mugilidae in the St. Lucia lake system. *Biol. J. Linn. Soc*., **8** (3).

Bogachik, T.A. 1969. Morphological adaptations of the digestive apparatus in some Labridae. In : *Biologia Morya* , **16** : 82-101, *Kiyev. Naukowa dumka* (In Russian).

Bose, K.C. and Islam, Md. Anisul 1986. Studies on the Morphohistology of the alimentary canal of *Anabas testudineus* (Bloch). *Proc. 73rd. Indian Sci. Congr., III* (vii) : 116.

Branch, G.M. 1966. Contribution to the functional morphology of fishes Part II. The feeding mechanism of *Syngnathus acus*. Lin. *Zool. Afr*., **2** : 69-89.

Bucke, D. 1971. The anatomy and histology of the alimentary tract of the carnivorous fish the pike, *Esox lucius* Lin. *J. Fish Biol*. **3** : 421-431.

Capanna, E., Catandella, S. and Honaco, G. 1974. The pharyngeal structure of mediterranean Mugilidae. *Monitore Zool. ital* (*N.S.*), t. **8**

Das, A.R. 1971. Anatomy, histology and histochemistry of the digestive system of *Ompok bimaculatus* (Bl.). *Proc. 58th. Indian Sci. Congr., III.* : 612-613.

Das, A.R. 1974. Anatomy, histology and histochemistry of the digestive system of a marine teleost, *Pisodonophis boro* (Ham.). *Proc. 61st. Indian Sci. Congr., III* : 70-71.

Das, A.R. 1982. Histomorphology and histochemistry of the digestive system of *Macrognathus aculeatus* (Bloch). *Proc. 69th Indian Sci. Congr.. III* (vii) . 85.

Das, A.R., Roy, K., Datta, S. and Bannerjee, S. 1974. Histomorphology of the digestive system of a freshwater teleost: *Heteropneustes fossilis* (Bloch). *Proc. 61st Indian Sci. Congr., III* : 71-72.

Das, S.M. 1968. Quantitative assessment of the gut contents of some freshwater fishes of North India. *Proc. 55th Indian Sci. Congr., III* : 513-514.

Das, S.M. and Moitra, S.K. 1955. Studies on the food of some fishes of Uttar Pradesh, Part I : The surface-feeders, the mid-feeders and the bottom-feeders. *Proc. Nat. Acad. Sci. Indi,* **25 B** (1-2) : 1-6.

Das, S.M. and Moitra, S.K. 1956a. Studies on the food of some common fishes of Uttar Pradesh, India, Part II: On the types of fish food and the variations in the relative length of the alimentary canal, with a description of the latter. *Ibid.,* **26 B** (4) : 213-223.

Das, S.M. and Moitra, S.K. 1956 b. Studies on the food of some common fishes of Uttar Pradesh, India, Part III : Comparative anatomy of the alimentary tract and its modifications in relation to their feeding habits. *Ibid.,* **26 B** (4) : 224-233.

Das, S.M. and Moitra, S.K. 1958. On the feeding types of fishes and the variations in the alimentary canal in relation to food. *J. Ichthyol, Moscow,* No. **10** : 29-40

Das, S.M. and Moitra, S.K. 1963. Studies on the food and feeding habits of some freshwater fishes of India, Part IV : A review on the food and feeding habits, with general conclusions. *Ichthyologica,* **2** (i-ii) : 107-115.

Das, S.M. and Nath, Surendra 1965. The comparative anatomy of the alimentary tract and its modifications in relation to food and feeding habits in some fishes of Jammu Province (India). *Ibid.,* **4** (1-2) : 63-78.

Das, S.M. and Nath, Surendra 1967. Comparative functional anatomy of the alimentary tract in some Kashmir fishes. *Kashmir Sci.,*,**4** : 38-53.

Das, S.M. and Nath, Surendra 1968. The comparative anatomy of the buccopharynx in some freshwater teleosts of Jammu and Kashmir (India). *Proc. 55th Indian Sci. Congr.*, III : 512.

Das, S.M. and Nath, Surendra 1973. On the guts of some herbivorous, omnivorous and carnivorous freshwater teleosts from Kashmir Valley. *Proc. 60th Indian Sci. Congr.,* III : 548.

Das, S.M. and Pathani, S.S. 1978. Studies on the biology of the Kumaon mahseer, *Tor putitora* (Ham) : Adaptation of the alimentary tract in relation to feeding habits, body-weight and body-length. *Indian J. Anim. Sci.,* **48** (6) : 461-465.

Das, S.M. and Subla, B.A. 1969. The mechanism of feeding in nine Kashmir fishes, with a comparative account of the standard mechanism in a herbivore, an omnivore and a carnivore. *Kashmir Sci.,* **6** (1-2) : 121-130.

Dasgupta, M. 1988. Adaptation of the alimentary canal in relation to feeding habits in *Boleophthalmus boddarti* (Pallas). *Proc. 57th. Indian Sci. Congr.,* III (vii) : 114.

Datta, N.C. and Das, M. 1982. Observations on the morphology of the alimentary tract of *Cynoglossus arel* (Schn.) in relation to its food and feeding habit. *Proc. 69th Indian Sci. Congr.* III (Vii) : 76-77.

Datta, N.C. and Dasgupta, M. 1986. Morphometrics of the filtering apparatus of *Oreochromis mossambica* Peters and *Catla catla* (Ham.) in relation to their food and feeding habits. *Proc. 73rd. Indian Sci. Congr.,* III (7) : 27-28.

De Groot, S.J. 1971. On the interrelationships between morphology of the alimentary tract, food and feeding behaviour in flat fishes (Pisces : Pleuronectiformes). *Neth. J. Sea Res.,* **5** : 121-196.

De Martini, E.E. 1969. A correlative study of the ecology and comparative feeding mechanism morphology of the Embiotocidae (surf perches) as evidence of the family's adaptive radiation into available ecological niches. *Wassman J. Biol.,* **27** (2) : 177-247.

DeWitt, F.A. and Cailliet, G.M. 1972. Feeding habits of two bristle-mouth fishes, *Cyclothone acclinidesn* and *C. signata* (Gonostomatidae). *Copeia,* 1972 (4) : 868-871.

Dexit, R.K. and Bist, J.S. 1972. Histomorphology of the alimentary canal of a hill-stream teleost, *Schizothorax richardsonii* (Gray & Hardw.) *Zool. Beitr.,* **18** : 321-328.

Gaur, R.D. and Jaish, S.C. 1970. Studies on the morphology and physiology of the alimentary canal of carnivorous fish, *Channa gachua* (Ham.). *Proc. natn. Acad. Sci. India,* **40B**(4) : 267-275.

Gaur, R.D. and Qureshi, N.A. and Irfan, N. 1975. Studies on the comparative histomorphology of the buccopharynx in relation to feeding habits in teleosts. *Proc. 62nd. Indian Sci. Congr.,* III (7) : 205.

Ghose, C.R., Ghosh, U. and Manna, B. 1982. Anatomy and histology of the digestive organs of *Ophicephalus punctatus* Bloch, *Proc. 69th. Indian Sci. Congr.,* III (vii) : 73-74.

Gosline, W.A. 1973. Considerations regarding the phylogeny of Cypriniformes fishes, with special reference to structures associated with feeding *Copeia,* 1973 : 761-776.

Grover, S.P. 1975. Food and feeding habits of Doon Valley fishes. *Proc. natn. Acad. Sci. India,* Annual No: (1975) : 83.

Gupta, O.P. 1971. Studies on the morphology, histology and the swallowing mechanism of the digestive tract of a carnivorous fish, *Xenentodon cancila* (Ham). *Okajimas Folia Anatomica Japonica,* **48** : 29-51.

Hynes, H.B.N. 1950. The food of freshwater sticklebacks (*Gasterosteus aculeatus* and *Pygosteus pungitius*), with a review of the methods used in studies of the food of fishes. *J. Anim. Ecol.*, **19** (1) : 36-58.

Ishiwata, N. 1969. Ecological studies on the feeding of fishes. *Bull. Japan. Soc. Scientific Fish*, **35** : 979-984.

Jacobshagen, E. 1911, 1913, 1915. Untersuchungen uber das Darmsystem der Fische und und Dipnoer I. III. *Jena. Z. Naturw.*, **47** : 528-568; **49** : 373-810; **53** : 445-556.

Kafuku, T. 1975. An ontogenetical study of intestinal coiling pattern in Indian Major carps. *Bull. Freshw. Fish. Res. Lab. Japan*, **27** (1).

Kamal, M.Y. 1967. Studies on the food and alimentary canal of Indian major carps-II : *Labeo rohita* (Ham.) and III : *Cirrhinus mrigala* (Ham.) *Indian J. Fish.*, **14** : 24-27.

Kamat, N.D. 1969. Algae as food of *Barbus conchonius* Ham. *Curr. Sci.*, **38** (2) : 50.

Kapoor, B.G., Smit, H. and Verighina, L.A. 1975. The alimentary canal and digestion in teleosts. *Adv. Mar. Biol.*, **13** : 109-239.

Karamchandani, S.J. and Mishra, D.N. 1978. Food and feeding habits of *Labeo boggut* (Sykes) from Kulgarhi Reservoir (Madhya Pradesh). *J. Bombay nat. Hist. Soc.*, **75** (2) : 389-396.

Kazansky, V.I. 1964. Specific differences in the construction of gill rakers in Cyprinidae. *Vopr. Ikhtiol.*, **4**: 45-60.

Keast, Allen and Webb, D. 1966. Mouth and bodyform relative to feeding ecology in the fish fauna of a small lake, Lake Opinicon, Ontario. *J. Fis. Res. Bd. Canada*, **25** (2) : 285-297.

Khalilov, F. Kh. 1969. *Materials on morphology and histochemistry of the digestive system in Teleosts.* Maktap, Alma Ata, 131 pp.

Khalilov, F.Kh., Inyushin, V.M. and Vorobjov, N.A. 1963. On morphology and histochemistry of the gut of fishes. *Izvestia Akad. Nauk. Kazakh. SSSR.* No. **2**, Ser. Biolog., 82-89.

Khanna, S.S. 1961. Alimentary canal in some teleostean fishes. *J. Zool. Soc. India*, **13** : 206-219.

Khanna, S.S. and Mehrotra, B.K. 1970. Histomorphology of the buccopharynx in relation to feeding habits in teleosts. *Proc. natn. Acad. Sci. India*, **40 B** (I) : 61-80.

Khanna, S.S. and Mehrotra, B.K. 1971. Morphology and histology of the teleostean intestine. *Anat. Anz.*, **129** : 1-18.

Kline, K. 1978. *Aspects of Digestion in Stomachless fishes*. Ph.D. Thesis, Univ. California, Davis, 78 pp.

Kozhina, E.S. 1958. Morphological peculiarities of the branchial apparatus of fish in relation to the mode of feeding. *Tr. Karel S.K. Filiala. Akad. Nauk SSR.*, **13** : 61-69.

Krayukhin, B.V. 1959. Effect of monotonous food on the digestive system of Carp. *Zoolog. Zhur.*, **33** (8).

Kryzhanovsky, S.G. 1948. Ecological groupings of fishes and regular features in their formation. *Trans. Pacific Inst. Fish., U.S.S.R.* **27**.

Kryzhanovsky, S.G. 1949. Ecological-morphological principles in the development of Cyprinid, Cobitidid and Silurid fishes. *Trans. Inst. Morf. zhiv.*, **1**

Kyle, H.M. 1926. *The biology of fishes.* London, Sedgwick & Jackson

Lagler, K.F., Bardach, J.E. and Miller, Robert R. 1962. *Ichthyology, the study of fishes.* John Wiley & Sons, New York

Lal, M.B. 1968. Studies on the anatomy and histology of the alimentary canal of a carp, *Tor putitora* (Ham.). *Proc. natn. Acad. Sci. India*, **B 38** : 127-136.

Lange, N.O. 1966. Structure and development of the pharyngeal teeth in roach, Caspian roach. In : *Morfo-ekol. Analiz. Rasvitiya Ryb* , Nauka (In Russian), pp. 163-177.

Liem, K.F. 1967. Functional morphology of the integumentary, respiratory and digestive systems of the synbranchoid fish, *Monopterus albbus. Copeia*, No. **2** : 375-388.

Majumdar, N.N. and Saxena, B.P. 1961. On the palatal organs of *Labeo dero* (Hamilton). *Proc. natn. Acad. Sci. India*, **31 B** : 109-116.

Matthes, H. 1963. A comparative study of the feeding mechanisms of some African Cyprinidae (Pisces; Cypriniformes). *Bijdr. Dierk.* **33** : 3-35.

Mehrotra, B.K. and Khanna, S.S. 1969. Histomorphology of the oesophagus and the stomach in some Indian teleosts with reference to their adaptational features. *Zoolog. Beitr.*, **15** : 375-391.

Mester, L. 1973. Comparative study of the buccopharyngeal cavity in some species of fishes belonging to the Cobitidae family. *Rev. Roumaine de biol., ser. Zool.*, vol. **18**.

Mohsin, S.M. 1962. Comparative morphology and histology of the alimentary canals in certain groups of Indian teleosts. *Acta Zool. Stockholm*, **43** : 79-135.

Moitra, S.K. and Bhowmik, M.L. 1967. Functional histology of the alimentary canal of the young *Catla catla* (Ham.), an omnivorous surface feeding fish of Indian freshwaters. *Vestnik. Ceskoslov. spol. Zool.* **31** : 41-50.

Moitra, S.K. and Ray, A.K. 1977. Morpho-histology of the alimentary canal of an Indian freshwater perch *Colisa fasciata* (Bloch) in relation to food and feeding habits. *Anat. Anz.*, **141** : 37-58.

Moitra, S.K. and Ray, A.K. 1979. Comparative morpho-histology and anatomy of the digestive system of two Indian freshwater perches *Ambassis nama* (Ham.) and *Ambassis ranga* (Ham.) in relation to their food and feeding habits. *Zool. Jahrb. Anat.*, **102**.

Moitra, S.K. and Sinha, G.M. 1971. Studies on the morpho-histology of the alimentary canal of a carp, *Chagunius chagunio* (Ham.) with reference to the nature of taste buds and mucous cells. *J. Inland Fish. Soc. India*, **3** : 44-56.

Mok, Hin-Kiu, 1980. Notes on the classification of actinopterygian intestinal patterns. *Japan. J. Ichthyol.*, **27** (1) : 29-40.

Montgomery, W.L. 1977. Diet and gut morphology in fishes, with special reference to the monkey-faced prickle-back, *Cebidichthys violaceus* (Stichaeidae : Blennioidei). *Copeia*, 1977 (1) : 178-182.

Mota, Alves M.I. and Sawaya, P. 1974. Aspectos to aparelho digestivo e da alimentagao de *Opisthonema oglinum* (Le Sueur) (Pisces, Clupeida *Arg. Cien. Mar.*, **14** (2).

Nair, N.B. and Sobhana, B. 1980. Food and feeding habits of *Puntius sarana subnasutus* (Val.). *Proc. Indian natn. Sci. Acad.*, **B46** (1) : 33-40.

Nath, Surendra 1972. On the comparative anatomy of the buccopharynx in some freshwater teleosts of Jammu and Kashmjr (India). *Proc. natn. Acad. Sci. India*, Annual No. (1972) : III

Nath, Surendra 1973. The comparative functional anatomy of the buccopharynx in four species of freshwater teleosts from Kashmir Valley. *Proc. 60th. Indian Sci. Congr., III* : 559.

Nath, Surendra 1976. On the incidence of stenophagy in *Glyptothorax kashmirensis* Hora (Siluriformes : Sisoridae). *Matsya*, **2** : 79.

Nath, Surendra 1978. On the food, feeding habits and alimentary tract of *Garra gotyla gotyla* (Gray) (Cypriniformes; Cyprinidae; Garrinae) *Ibid.*, **4** : 75-76.

Nath, Surendra 1981. On the pharyngeal teeth of *Noemacheilus stoliczkae* (Steind.) (Cypriniformes; Cobitidae; Noemacheilinae), the high altitude cobitid loach of Kashmir. *J. Zool. Soc. India*, **33** (1 &2) : 149-150.

Nath, Surendra 1982. On the food, feeding habits and alimentary canal of *Rasbora daniconius* (Ham.) (Cypriniformes; Cyprinidae; Rasborinae) *Matsya*, **8** : 64-67.

Nath, Surendra 1984. On the food, feeding habits and alimentary canal of a hill-stream teleost-*Barilius vagra* (Ham.) (Cypriniformes; Cyprinidae; Bariliinae). *J. Zool. Soc. India*, **36** (1 & 2) : 121-124.

Nath, Surendra 1985. On the food, feeding habits and alimentary canal of *Glyptothorax lineatus* (Day) (Siluriformes; Sisoridae)-A torrential-stream teleost from Poonch Valley (Jammu and Kashmir State). *Matsya*, **II** : 89-92.

Nath, Surendra 1986. Observations on the general ecology of fishes of Poonch Valley (Jammu and Kashmir State) India. *Livestock Adv.* **II** (12) : 31-36.

Nath, Surendra 1988. On the food, feeding habits and alimentary canal of two Cobitid loaches of the genus *Noemacheilus* van Hasselt from Poonch Valley (J. & K. State). *Rec. Adv. Fish. Ecol. Limn. Eco. Conserv.*, **I** : 69-72.

Nath, Surendra 1989a. Studies on the guts of some herbivorous, omnivorous and carnivorous freshwater teleosts from Jammu and Kashmir (India) Part I : Anatomy. *J. Zool. Soc. India*, **40** (1-2) : 1-12.

Nath, Surendra 1989b. Studies on Systematics, Ichthyogeography and Bio-ecology of Fishes of Jammu Province (Jammu & Kashmir State) India. Ph. D. Thesis, Kanpur University, Kanpur.

Nath, Surendra 1992. Studies on the bioecology of fishes of Jammu Province (Jammu and Kashmir State) India, Part II : Feeding Ecology. *Proc. natn. Acad. Sci. India*, Annual No. 1992.

Nath, Surendra and Das, S.M. 1970. On the comparative anatomy of the buccopharynx in some freshwater teleosts of Jammu and Kashmir (India). *Proc. 57th. Indian Sci. Congr., III* : 387.

Nelson, G.J. 1972. Observations on the gut of the Osteoglossomorpha. *Copeia*, 1972 (2) : 325-329.

Nikolskaya, N.G. and Verighina, I.A. 1974. Peculiarities of the structure of feeding and digestive organs in three flatfish species of the White Sea associated with the nature of food consumed by them. *Vop. Ikhtiol.*, **14** : 117-128 (In Russian)

Nikolsky, G.V. 1963. *The Ecology of Fishes*. 352 pp. Academic Press, N.Y.

Nikolsky, G.V. 1974. *Ekologia Ryb*. 367 pp, Vysshaya Schkola, Moscow.

Nirmala, K. and Sethi, R.P. 1970. Morphology and histology of the alimentary tract of a major Indian carp, *Cirrhina mrigala* (Ham.). *Proc. 57th. Indian Sci. Congr., III* : 388.

Oppel, A. 1896-1897, 1900. *Lehrbuch der Vergleichenden mikroskopischen Anatomie der Wirbelthiere*. Bd. **I**. S. 28 et seq., Bd. **2**. S. 7 et seq., Bd. **3**.

Pandey, K.C. and Pandey, A.K. 1972. On the histomorphology of freshwater carnivorous fish *Nandus nandus*—The alimentary canal. *Proc. natn. Acad. Sci. India*, Annual No. 1972 : II.

Pathani, S.S. 1979. Studies on the Ecology and Biology of Kumaon mahseers Tor tor and T. putitora (Ham.). Ph.D. Thesis, Kumaun University, Nainital.

Pictet, A. 1909. Contribution a l'etude histologique due tube digestif des poissons cyprinoides. *Rev. Suisse de Zool.*, T. **7** : 1-78.

Pietsch, T.W. 1978. The feeding mechanism of *Stylophorus chordatus* (Teleostei: Lampridiformes) : functional and ecological implications. *Copeia*, **(2)** : 255-262.

Pillay, T.V.R. 1951. A critique of the methods of the study of the food of fishes. *J. Zool. Soc. India*, **4** (2) : 185-196.

Polyakov, G.D. 1966. Variation in the length of intestine of the Carp in relation to the feeding conditions. *Zool. Zhur.*, **15** (8).

Popta, 1900. *Ann. Sci. Nat. Zool.*, T. **12** : 139-216.

Rahimullah, M. 1945. A comparative study of the morphology, histology and probable functions of the pyloric caeca in Indian fishes, together with a discussion on their homology. *Proc. Indian Acad. Sci.*, **B21** : I-37.

Rao, K.H., Chakravarty, M.S. and Shyamasundari, K. 1980. Comparative morphology and organisation of the alimentary canal in the Lizard fishes : *Saurida tumbil* (Bloch), *Saurida undosquamis* (Richardson) and *Trachynocephalus myops* (Schneider) (Family : Synodidae). *Proc. natn. Acad. Sci. India*, **50 B** (4) : 211-217.

Rastogi, R.K. 1966. The morphology and histology of the alimentary canal of *Heteropneustes fossilis* (Bloch). *Agra Univ. J. Res.*, **16** : 63-72.

Reshetnikov, Yu 1961. On the connection between the amount of gill rakers and the nature of feeding in Chars of the genus *Salvelinus*. *Zoolog. Zhur.*, **40** : 1574-1576. (In Russian)

Saxena, O.P. 1980. Relationship of buccopharynx with the food and feeding habits of *Labeo dero* (Ham.). *Proc. Indian natn. Sci. Acad.*, **B46** (I) : 41-47.

Schmitz, E.H. and Baker, C.D. 1969. Digestive anatomy of the gizzard shad *Dorosoma cepedianum* and the threadfin shad, *D. petenese*. *Trans. Amer. Micros. Soc.*, **88** : 525-546.

Sehgal, P. 1966. Anatomy and histology of the alimentary canal of *Labeo calbasu* (Ham). *Res. Bull. Punjab Univ.*, **17** : 257-266.

Sehgal, P. and Salaria, J. 1970. Functional anatomy and histology of the digestive organs of *Cirrhina mrigala* (C. & V.). *Proc. natn, Acad. Sci. India*, **40 B** (4) : 212-222.

Sharma, R.S. 1984. Studies on the predatory habits of two teleost fishes, *Rasbora daniconius* (Ham.) and *Barbus ticto* (Ham.) in the zoological Laboratory. *Proc. 71st. Indian Sci. Congr.*, *III* (7) : 124.

Sharma, V.K. and Tandon, K.K. 1982. Food and feeding habits of some fishes of Kangra Valley, H.P. *J. Zool. Soc. India*, **34** : 86-95.

Shpet, G.I., Karitonova, N.N. and Bakunenko, L.A. 1962. On the morphology of the gill apparatus in Crucian carp and Common Carp in relation to differences in their feeding. *Vopr. Ekolog.*, **5**.

Shuljak, G.S. 1968. On the morphology of the gut of Silver carp. *Gidrobiol. Zhur.*, **4** : 30-38.

Shuljak, G.S. 1968 b. On the morphology of goldfish. *Ibid.*, **4** (4).

Sibbing, F.A. 1976. Pharyngeal mastication in *Cyprinus carpio* (Lin.). In : *Acts of the 2nd. European Ichthyological Congress*.

Singh, K.P. and Samuel, P. 1981. Food, feeding habits and gut contents of *Anabas testudineus* (bloch). *Matsya*, **7** : 96-97.

Singh, R. 1966. Morpho-histological studies on the alimentary canal of *Bagarius bagarius* Ham. *Agra Univ. J. Res. (Sci.)*, **15** ; 69-82.

Singh, R. 1967. Studies on the morphology and histology of alimentary canal of *Ophicephalus marulius*. *Ibid.*, **16** : 27-38.

Singh, R. 1967 b. Studies on the anatomy and histology of the alimentary canal of some freshwater teleosts in relation to their feeding habits. *Ibid.*, **16** : 165-168.

Singh, S.P. 1974. The morphology and anatomy of the alimentary canal of a silurid fish *Ompok bimaculatus* (Bl.) *Proc. 61st. Indian Sci. Congr.*, III (viii) : 93.

Sinha, G.M. 1973. The comparative morphology and anatomy of the alimentary canal of four Indian freshwater carps during the different life-history stages in relation to food and feeding habits. *Proc. 60th Indian Sci. Congr.*, III : 548-549.

Sinha, G.M. and Moitra, S.K. 1975. Morpho-histology of the intestine in a freshwater major carp, *Cirrhinus mrigala* (Ham.) during the different stages in relation to food and feeding habits. *Anat. Anz.* Bd. 137.

Sriwastava, V.M.S. 1970. Functional anatomy of the digestive organs of a freshwater fish *Rhynchobdella aculeata* (Ham.) I A. Morphology. *Vestnik. Ceskoslov. Spol. Zool* , **31** : 129-135.

Stein, R.A. and Kitchell, J.F. 1975. Selective predation by a carp *Cyprinus carpio* (Lin.) on benthonic molluscs in Skadar Lake, Yugoslavia. *J. Fish. Biol.*, **7** (2) : 391-399.

Stroband, H.W. 1977. Growth and diet dependent structural adaptations of the digestive tract in juvenile grass carp (*Ctenopharyngodon idella* Val.). *Ibid.*, **II**.

Suvorova, E.G. and Treschuk, L.I. 1973. The morphology and histochemistry of the gastro-intestinal tract of the White Baical grayling *Thymallus arcticus baicalensis* Dyb. *Vopr. Ikhtiol.*, **13** : 523-534.

Suyehiro, Y. 1942. A study on the digestive system and feeding habits of fish *Japan. J. Zool.*, **IO** : 1-303.

Svetovidov, A.N. 1934. On the dependence between the feeding habits and the number of pyloric caeca in fishes. *Doklady Akademii Nauk SSSR*, **3** : 67-72 (In Russian).

Svetovidov, A.N. and Skorzowa, T.A. 1968. The structure and functional significance of pharyngeal organs in Clupeoid fishes. *Trudy Zoologicheskogo Inst. Akad. Nauk SSSR, Leningard*, **46** : 169-189.

Szarski, H. 1956. Causes of the absence of a stomach in Cyprinidae. *Bull. Acad. Polonaise Sci.*, Cl. II. *Zoology*, **4** : 155-156.

Tandon, K.K and Goswami, S.C. 1968. A comparative study of the digestive system of *Channa* species. *Res. Bull.* (N.S.) *Punjab Univ.*, **19** : 13-3.

Tandon, K.K., Johal, M.S. and Dhawan, S.K. 1972. Food and feeding habits of some fishes of the Punjab. *Ibid.*, **23** (1-2) : 27.

Trewisan, P. 1979. Histomorphological and histochemical research on the digestive tract of freshwater grass carp *Ctenopharyngodon idella* (Cypriniformes). *Ant. Anz.*, Bd. **45**.

Vaidya, V. 1966. A description of the buccopharyngeal region in *Nandus nandus* Ham. *Proc. 53rd. Indian Sci. Congr.*, III : 370.

Valatour, M. 1961. Recherches sur les glandes gastriques et les tuniques musculaires due tube digest if dans les poissons, osseaux et les batraciens. *Ann. Sci. Nat., Zool.*, **4**, sern. T. **16** : 219-285.

Verighina, I.A. 1967. The structure of the alimentary tract in plant-eating fishes of the Cichlidae family. *Vopr. Ikhtiol.*, **7** : 362-371.

Verighina, I.A. 1969. The structure of the digestive tract of *Varicorhynus capoeta sevangi* (Filippi). *Ibid.*, **9** : 539-545.

Verighina, I.A. 1969b. Ecologo-morphological peculiarities of the alimentary tract of some Cypriniformes. In : "*Itogi nauki, Zoology, 1968.*" pp. 79-109. *Institut nauchanoi inforomatsii Akademii Nauk.*

Verighina, I.A., 1971. The structure of the digestive tract of the Volga undermouth *Chondrostoma nasus variabile* Jak. *Ibid., II* : 311-318.

Verighina, I.A. 1977. Structural features of the alimentary canal in connection with the lack of a stomach in the Black Sea blenny, *Blennius sanguinolentus*. *J. Ichthyol., Moscow*, **17** (6) : 964-968.

Verighina, I.A. 1979. Structure of the alimentary tract of the deepsea fish *Rouleina attrita* (Vaillant) (Alepocephalidae, Salmoniformes). *Bull. MOIP*, **85** (4) : 54-57.

Verighina, I.A. and Medani, J.I. 1968. The structure of the digestive tract in *Distichodus niloticus* (Lin). and *D. rostratus* (Gunther) in relation to feeding habits. *Vopr. Ikhtiol.*, **8** : 710-721.

Verighina, I.A. and Savvaitova, K.A. 1974. The structure of the alimentary tract in some chars (g. *Salvelinus*) from sympatric populations of Lake Azabachje (K̆amchatka). *Ibid.*, **14** : 466-476.

Verighina, I.A. and Zoldasova, I.M. 1982. *Ecologo-morphological peculiarities of the digestive system of Teleostei.* "Fan", Tashkent.

Verma, S.R., Dalela, R.C., et al. 1974. A study of the alimentary canal and physiology of digestion in *Gadusia. Zool. Jahrb.*, **92** : 445-452.

Verma, S.R., Tyagi, M.P. and Dalela, R.C. 1974. Morphological variations in the stomach of few teleost fishes in relation to food and feeding habits. *Gegenbaur's morphol. Jahrb.*, **120** : 244-253.

Weinberg, S. 1976. Morphology of the intestine of the Goldfish (*Carassius auratus*). *Bijidr. Dierke.*, **46**(1).

Weisel, G.F. 1973. Anatomy and histology of the digestive system of the paddlefish (*Polyodon spathula*). *J. Morph.*, **140** : 243-256.

Western, J.R.H. 1969. Studies on the diet, feeding mechanisms and alimentary tract in two closely-related teleosts, the freshwater *Cottus gobio* Lin. and the marine *Parenophryz bublis* Euphrasen. *Acta Zool. Stockholm*, **50** : 185-205.

Witte, F. and Barel, C.D.N. 1976. The comparative functional morphology of the pharyngeal jaw apparatus of piscivorous and intra-pharyngeal mollusc-crushing *Haplochromis* sps. *Acts. 2nd European Ichth Congr.*

Zander, E.. 1903. Studien uber das kiemenfilter bei susswasserfiscehen. *Zeitsch. fur Wiss. Zool.*, Bd. **75** : 233-258.

Zander, E. 1906. Das kiemenfilter der Teleostier, eine morphologische studie. *Ibid.*, Bd. **84** : 619-713.

Zobairi, A.R.K. and Choodamanui, N.V. 1948. Food, feeding habits and rate of growth of the Mirror Carp—*Cyprinus carpio* Lin. *Proc. 35th. Indian Sci. Congr.*, III : 202-203.

Rec. Adv. Fish Ecol. Limn. Eco-conserv.
IV : 107-113

11

The Vanishing High Altitude Animals of the Himalayas

S.M. DAS

The Red Data book records 600 species of animals threatened with extinction all over the world today. That 106 species of mammals and 139 species of birds became extinct during the past 2000 years of human civilisation is also a recorded fact. During the past 100 years alone as many as 70 species of animals have become extinct; and over 50 species of animals have vanished from our earth during the current century.

Among the fast-vanishing wildlife in the world today, none is being decimated or disappearing faster than the wildlife of the Himalayas. In Jim Corbett's heyday, the Himalayas harboured large number of tigers, as also barking deer, tree leopard, Tibetan antelope, Tibetan gazelle, wild boar, langur, golden-, leopard-, and civet-cats, and marten up to 10,000 feet; between 6000 and 14,000 feet many wild animals were common : snow-, and clouded-leopards, musk deer, woolly wolf, brown bear, panda, cat bear, stone marten, hungul or Kashmir stag, chiru or Tibetan antelope, goral, tahr (Kashmir Sikku), ibex, markhor, bharal, nayan (Tibetan sheep), shapu, urial, flying squirrel, mouse hare (up to 17,000 feet), marmot, limbless lizard, toad lizard (Phrynocephalus), Himalayan toad (*Sentiger*) and high altitude skinks (*Leiolopisa* and *Lygosoma*).

The Himalayan biotope is a sub-region of the Oriental Indian region with palaearctic affinities. Almost one-third of the Kashmir fauna is palaearctic in origin and about a quarter of Nepal's fauna also. Palaearctic incursions have also occurred in U P Himalayas (Uttarkhand Himalayas), where the fauna is more or less distinct from that of the rest of India. The large altitude range results in a characteristic zonation of animals and much more so of plants on which the animals are dependent. The main zones are: forest zone, moist temperate, arid plateau, and alpine forest. Forest zone can be divided into four sub-zones:

1. Tropical deciduous forest which comprises the Doons or foot-hills and lower valleys between 500-2000 ft. (150-600 m.), where the dense forests consist mainly of sal, shisham, silk-cotton and giant bamboo. The tropical forests rise higher in the eastern than in the western Himalayas and tiger, elephant, gaur, sambhar, and muntjac are the typical large animals. Annual rainfall is 150-250 cm.
2. The Terai-Bhabar Zone between 1000 and 5000 feet (300-1500 m.) which consists of a sloping low-lying country adjoining the foothills, and the

home of water buffalo, chital, tree cat, hog deer, swamp deer, blackbuck, barasingha, chousingha, nilgai, Indian fox, jackal and porcupine. Annual rainfall is 250-300 cm.

3. The Low hills between 5000 and 6000 feet (1500-1800 m.) which are covered with dark and gloomy rain forests of oak, deodar, cypress, pine (*Pinus roxbergi*), birches, magnolias, laurels and epiphytes, and have the typical fauna of Kumaon hills such as black-faced langur, sloth bear, honey badger, yellow-bellied marten, Himalayan otter, stone marten, high range tiger, leopard, caracal, spotted civet, palm civet, hill fox, spotted flying squirrel, black bear, occasional low range tiger, etc. Annual rainfall is from 250 to 375 cm.
4. Medium wet mountains between 6000 and 9000 feet (1800-2700 m.) where birches, pure stands of rhododendrons, juniper, yew, dwarf bamboos, some dwarf oaks and cypresses abound up to the tree line at 4000 m., and where the typical animals are hog badgers, crestless porcupine, goral, with serow in the higher reaches. Rainfall here is medium, from 125 to 250 cm. per year.

Moist Temperate Zone

The moist temperate zone is more marked in the western Himalayas, 2000 to 4000 m. of Kashmir, Hazara, Chitral, Gilgit, and Hunza, *Abies, Cedrus, Picea, Tonga* and *Pinus excelsa* are the typical conifers; *Quercus*, laurels, *Acer, Prunus, Ulmus, Betula* and some epiphytes also occur. The rare animals include giant flying squirrel, ibex, hangul (Kashmir stag), musk deer, Himalayan marten, Tibetan lynx, black bear and panda cat bear. Rainfall ranges from 75-127 cm in a year, but snowfalls are common.

Arid Plateau

The arid plateau is typified by Ladakh in J. & K. State and Tibet (now in China) and also some high altitude regions of Uttarkhand (3000-5000 m.). Many parts of Ladakh are cold deserts but other plateaus are oases of grass, with stunted birches, junipers and berries, and a few beautiful meadows where Gujjars graze their sheep and cattle in summer. Here pockets of goral, bharal, ibex, markhor, takin, chus and Tahr are found. The average rainfall is 8-25 cm in a year.

Alpine Forests

Stands of Alpine forests are scattered in regions with both rainfall (25-75 cm) and snow, at a height of 4000-5000 m. They consist of crooked trees, large shrubs with some conifer overwood, as also dwarf rhododendrons, interspersed with valleys of flowers and Alpine meadows where gentians, primulas, iris, saxifrages, geraniums, thyme and asters bloom. Himalayan larch (*Larix*), *Abies, Betula* are the typical trees, epiphytic mosses and lichens. These forests stretch only up to the snowline, which varies from mountain to mountain. At about 5000 m. the vegetation changes to alpine scrub and grasslands; above that are rugged empty regions of rock, ice and glaciers where only guggal and mamla grass grow in the short summers.

From south to north the mountain region is divisible into four geological zones : The outer zone, the Siwalik hills: which includes the Bhabar and the longitudinal duns and averages about 50 km. across, with an elevation of 5000 to 1250 m. The middle, lesser or sub-Himalayan zone : with a width of about 80 km. and an elevation up to 3000 m. The greater or Central Himalayan zone : the zone of the loftiest peaks over 3000 m., extending south from the Indo-Tibetan water parting ridge, with a width of 30 to 50 km. The trans-Himalayan zone : which includes the area between Central Himalayas and the Indo-Tibetan water parting ridge. The terrain mainly comprises steep spurs and deep gorges.

CLIMATE OF UTTARKHAND HIMALAYAS AND HUMAN HABITATIONS AFFECTING WILDLIFE

The climate ranges from tropical in Bhabar and low valleys to extreme arctic in the snow and ice-bound areas. The variation in exposure to sunlight and to rain-bearing winds produce intricate patterns of local climate.

The south-west monsoon from June to September accounts for 80 per cent of the precipitation (rain plus snowfall). The Duns and Bhabar receive about 150 to 300 cm, the fall gradually diminishing from east to west. To the north, except for valleys such as Nainital Bhowali-Bhimtal, rainfall rapidly decreases and practically ceases beyond the first snow range. October and November are clear dry months, but temperature fluctuations from cool to cold and dew and frost are common. Winter rains (December to February) occasionally occur, but usually there are snowfalls even down to 2000 m. March and April (Spring season) are often marred by high gales, hailstorms and occasional snow.

In the Central and trans-Himalayan zone (above 3000 m.) snow falls from mid-November to mid-April, and everything is buried under snow. Most small mammals hibernate or wintersleep, and even man is rarely seen out of doors. But this is the time that snow-leopard, brown bear and muskdeer descend to 2000 metres in search of food and are easily shot and killed. They also descend with the decreasing snowline into the gorges, glaciers and meadows down to 2000 metres.

Human habitations range from 1000 to 4000 metres, occasionally higher. Ribbon-like fields have been carved out on the gentler mountain slopes and in river valleys. Even highland areas are undergoing population explosion (with 4-8 children per couple), wherever agrarian and orchard land is available. In winter these people migrate to lower elevations with their livestock. In summer sheep and goats are taken into high alpine pastures (up to 5000 m.). This is when poachers, shikaries and even urban people ravage and decimate high altitude wildlife for food and for the fur-, skin-, horn- and head-trophy trade.

RARE MAMMALS

There are 30 or more species endangered in Himalayas today. The alpine forest is the home-land of many rare animals of the Himalayas :

Snow leopard	:	*Uncia (Panthera) uncia*
Himalayan Marten	:	*Mustela sibirica hodgsoni*
Tibetan sheep (Nayan)	:	*Ovis ammon hodgsoni*
Ibex	:	*Capra ibex sibirica*
Yak	:	*Bos grunniens mutus*
Tibetan wolf (woolly)	:	*Canis laniger*
Tibetan gazelle	:	*Procapra picticaudata picticaudata*
Bharal	:	*Pseudois nayaur nayaur*
Tibetan Antelope	:	*Pantholops hodgsoni*
Himalayan wild dog	:	*Cuon alpinus ladaki*
Himalayan leopard	:	*Panthera pardus millardi*
Clouded leopard	:	*Neofelis nebulosa*
Kashmir Otter	:	*Lutra perspicillata*
Ermine (Himalayan)	;	*Mustela erminea ferghanae*
Himalayan Stoat	:	*Nemorhaedus goral goral*
Black bear	:	*Selenarctos thibetanus laniger*
Hangul	:	*Cervus elaphus hangul*
Pigmy hog	:	*Sus scrofa cristatus*
Tahr or Jharal	:	*Hemitragus jemalhicus jemlahicus*
Himalayan Mouse hare	;	*Ochotona royelei royelei*
Himalayan Vole	:	*Alticola roylei roylei (caudatus, glacialis, blanfordi).*
Tibetan red marmot	:	*Marmota bobak himalayana*
High altitude flying Squirrel	:	*Hylopetes albiniger*
Clawless Otter	:	*Aonyx cinerea concolor*
Brown bear	:	*Ursus arctos isabellinus*
Musk deer	:	*Moschus moschiferus moschiferus*
Giant flying Squirrel	:	*Callosciurus pygerythrus*
Red flying Squirrel	:	*Petaurista petaurists albiventer*
Kashmir flying Squirrel	:	*Eupetaurus cinereus*
Red Fox	:	*Canis vulpes montana*

The following large Himalayan mammals require protection today:

Musk deer (*Moschus moschiferus moschiferus*) : are extensively and illegally shot in Kashmir and Uttarkhand for the musk pod, each one of which fetches over Rs. 1,000. Strict enforcement of the law and the establishment of musk deer farms are essential if this rare animal is not to vanish from the Himalayas (Altitude 2000 - 5000 m.)

Flying squirrels (Kashmir : *Eupetaurus cinereus, Hylopetes fimbriatus,* and Kumaon : *Petaurista petaurista albiventer*) (1700 - 4000 m.), *Hylopetes* (1830 - 36000 m.) : are systematically hunted for their thick fur, both in Kashmir and Uttarkhand. Once abundant in Garhwal and Kumaon, they are now extremely rare. The hoary-bellied Himalayan squirrel *Callosciurus pygerythrus* (150 to 525 m.) occurs lower.

Bharal or blue sheep (*Pseudois nayaur nayaur*) : much prized for both meat and skin, is easier to shoot than the ibex or markhor (2000 - 3500 m.).

Mouse hare (*Ochotona roylei roylei*) has been so severely hunted that it is now hardly seen anywhere (3400 - 4300 m.).

Tibetan tahr or jharal (*Hemitragus jemlahicus jemlahicus*) has shared the Bharal's fate for the same reasons (3050 m. - 3660 m.).

Snow leopard (*Panthera uncia*) has almost disappeared from Kashmir, where it was once very abundant. The beautiful single fur coat sells for anything up to Rs.5,000. The author has been dozens of such skins with furriers in Kashmir. Unless this beautiful animal is preserved in high altitude biosphere reserves or parks, one more magnificent animal will disappear from the Himalayas (3660 - 3695 m.).

The panda cat bear (*Ailurus fulgens fulgens*), which can be easily tamed or even kept in enclosures, is now strictly protected (Above 1525 m.).

Brown bearn (*Ursus arctos isabellinus*) is now a rarity, as hill men go higher and higher in search of new forests for wood and pastures for cattle. The golden fur is also highly priced (2500 - 4500 m.).

Tibetan Gazelle (*Procapra picticaudata picticaudata*) is a cousin of the *Chinkara* of lower altitudes. It is the only gazelle of the Uttarakhand Himalayas. It is a shy animal and in human inhabited areas, numbers have nevertheless been decimated for meat and skin (2500 - 5000 m.).

Chiru or Tibetan antelope (*Pantholops hodgsoni*) is easier to shoot than the mountain goats or gazelles and is also endangered (2000 - 4500 m.).

Markhor (*Capra falconeri falconeri cashmirensis* of the western Himalayas, is now very rare and needs immediate conservation measures. It may weigh 100 kilos and is extensively shot for meat and skin (1500 - 4500 m.).

Woolly wolf (*Canis laniger*) has not been seen in Uttarkhand and Ladakh for some time now, and is probably extinct (3000 - 5000 m.).

Kashmir stag (*Cervus elaphus hangul*), now very rare (250 heads), extends from Kishenganga upper valley to Kishtwar and parts of Pir Panjal range (2000 - 3000 m.).

Himalayan ibex (*Capra ibex sibirica*) has a much-prized winter coat, the undercoat yielding the famous *Pashmina* that is naturally shed after winter; it inhabits precipitous high mountains in Kashmir, Ladakh, and western Himalayas (2500 - 5000 m.).

Red Fox (*Canis vulpes montana*) of Kashmir, has been decimated for its fur (2000 - 4000 m.).

Tibetan Lynx (*Felis lynx isabellinus*), highly prized for its silvery fur, was once abundant in Ladakh (4000 m.) and Garhwal (above 3000 m.), but is now rare.

Tibetan sheep or Nayan (*Ovis ammon hodgsoni*) is now extremely rare in the Himalayas. (3000 - 4575 m.).

THE VANISHING BIRDS

The endangered birds of the high Himalayas are :

Monal Pheasant	:	*Lophophorus impejanus*
Himalayan Snowcock	:	*Tetraogallus himalayensis*
Snow pigeon	:	*Columba leuconota*
The Chough	.	*Pyrrhocorax graculus*

Cheer Pheasant	:	*Catreus wallichii*
White-tailed Nutthatch	:	*Sitta himalayensis*
Himalayan Whistling Thrush	:	*Myophoneous temincki*
Himalayan Greenfinch	:	*Hypocanthis spinoides*
White-capped Red start	:	*Charmarrornis leucocephalus*
Spotted forktail	:	*Enieurus maculatus*
Striped laughing Thrush	:	*Trochalopteron lineatum*
Red-headed Tit	:	*Lophophanes melanoplus*
Green-backed Tit	:	*Parus indicus monticalus*
Himalayan Red-crowned Jay	:	*Garrulus bispecularis*
Himalayan Black-throated Jay	:	*Garrulus lanceolatus*
Himalayan Eagle	:	*Aquila chryseatus*
Himalayan flacon	:	*Falco peregrinaton*
Himalayan Chakor	:	*Alectoris graeca chakor*
Himalayan Vulture	:	*Vultura monachus*
Rufus babbler	:	*Argea subrufa*
Snow pigeon	:	*Columba leuconota*

HIGH ALTITUDE ANIMAL PHYSIOLOGY

All high-altitude animals have morphological as well as physiological adaptations seen mainly in the size of body and limbs, changes in blood composition, skin texture and colour, length of body hair (fur) and feather abundance, changes in lung capacity and blood supply and other minor adaptations.

In extreme cold conditions a bird or mammal has an advantage when its body surface is small in proportion to its size (Bergman's Law); the larger the size, the smaller its surface in proportion to mass. This is because mass increases as the cube of linear dimensions, and the surface radiates only as its square. Thus, larger bodies radiate less heat in high altitude cold. Bears are larger in cold Alaska than further south in U.S.A.; deer, fox, wolf, wild bear, goat and hare increase in size with the altitude; antelope on high Himalayas are more compact and much heavier than African antelopes. The proportionately smaller body surface of larger animals in high altitudes is also an effective protection against ultra violet rays, which are intense in high altitudes and are harmful to body cell-function, and may cause cancer.

Allen's law states that in colder region animals tend to have shorter protruding body parts—legs, ears, tail—for less heat loss by radiation in acute cold. This applies to almost all high Himalayan mammals and to some extent for man as well.

According to Gloger's law skin colour tends to be lighter in cold dry climates than in warm moist ones. Dark skin pigments prevent the sun-rays from penetrating the body, and prevent sunburn. Even man in cold high regions is susceptible to sunburn and skin cancer.

The blood also undergoes changes which affect the number and size of red blood corpuscles (RBCs) and also the haemoglobin content. Because there is less oxygen, high-altitude animals have more RBCs, and they are smaller because small RBCs present a larger surface are a (functional area) for exchange of gases and absorption of oxygen. People returning from high altitudes usually had red cheeks due to the production of more RBCs, and also due to contraction of the spleen also as a result of oxygen lack. Moreover, the coefficient of absorption of oxygen by the more abundant haemoglobin is also higher. High altitude animals produce more oxygen and more heat calories than low altitude animals; their heart action is stronger and more rapid, and the flow of blood through lungs (and body) can be twice as high—for example in snow leopard, musk deer, Tibetan sheep. The RBC count of a Ladakhi is above 8000, while in the plains man it is about 6000.

HIGH ALTITUDE WILDLIFE PRESERVATION

Few wildlife sanctuaries, biosphere reserves and national parks have been established in the high Himalayas, and there is urgent need for their establishment. Also, breeding centres for the vanishing species should be started, as has been done for the crocodile, rhinoceros and some birds. It is also urgent to stop the large export of Himalayan wildlife. The forests and high altitude wildlife should not only be included in the concurrent Central list (as the Indian Board of Wildlife proposed in 1976), but all matters concerning wildlife preservation must be transferred form the State list to the concurrent Central list urgently to ensure effective central guidance and control. The younger generation in the hill universities should be educated about their wildlife and the immediate urgency of conservation of the vanishing species of Himalayan animals.

REFERENCES

Atkinson, E.T. 1884, Fauna of the Himalayas. *Himalayan Gazetteer, I.*

Biswas, B. and Khajuria, H. 1995. Zoological results of the "Daily Mail" Himalayan Expedition, 1954. *Proc. Zool. Soc.* Calcutta, 9 : 229-253.

Biswas, B. and Tiwari, K.K. 1966. Taxonomy and distribution of Indian Rodents. *Indian Rodent Symposium*, Calcutta.

Bourliere, F. 1955. *Mammals of the World.* George Harrap, London.

Ellerman, J.R. and Morrison-Scott, T.C.S. 1955. Checklist of Palaearctic and Indian Mammals. *Brit. Mus. Nat. Hist.*, London

Ellerman, J.R. 1963. *The Fauna of India, Mammalia,* Vol. 3. *Rodentia.* Govt. of India Publication, Calcutta.

Finn, F. 1929. *Sterndale's Mammalia of India.* Thacker Spink & Co., Calcutta, 347 pp.

Jerdon, T.C. 1867. *The Mammals of India.* Roorkee, 319 pp.

Majupuria, T.C. 1986. *Wildlife Wealth of India.* Tecpress, Bangkok.

Nath, Surendra, 1993. A checklist of Mammals of Jammu and Kashmir State (India) with remarks on the Mammals geography of the State. *Livestock Adv.*, 18 (5) : 6-13.

Prater, S.H. 1965. *The Book of Indian Animals.* Bombay Nat. Hist. Soc. Bombay, 324 pp.

Sanderson, I.T. 1955. *Living Mammals of the World.* Hamish Hamilton, London.

Walker, E.P. 1964. *Mammals of the World.* John Hopkins Press, Baltimore, U.S.A., pp. 1500.

Rec. Adv. Fish Ecol. Limn. Eco-conserv.
IV : 114-115

12

Metallic Element Composition of *Pomphorynchus kashmirensis* Datta, 1936—An Example of Calcium Absence

M.K. RAINA, P.L. KOUL and K. NARAYAN

In the present investigation atomic absorption spectro-photometric analysis of *Pomphorhynchus kashmirensis* was conducted and interestingly the absence of calcium was revealed.

Live specimens of *P. kashmirensis* were collected from the intestine of *Schizothorax niger*. The parasites were washed thoroughly with distilled water and placed on a piece of filter paper to remove water. After weighing, the parasites were digested in con. nitric acid. The solution was diluted with ion-free water and subjected to analysis on atomic absorption spectrophotometer with the following parameters for various elements (gas pressure, air pressure and monochrometer band pass being 1 litre, 5 litres and 0.4 nm respectively for all the elements).

Element	*Tube current (mA)*	*Wave length (nm)*
Mg	3	285.21
Ca	8	422.76
Mo	12	313.26
Mn	10	279.48
Fe	12	248.33
Co	12	240.73
Ni	12	232.00
Cu	4.5	324.75
Ag	4	328.07
Zn	8	213.86
Hg	5	253.65
Pb	5	217.00

The readings taken were calibrated with the graphs obtained for standards.

The following elements were found to be absent :
Ca, Hg, Pb, Ni, Co, Ag and Mo.

Spectrophotometric readings for other elements were as under : Mg. 0.109, Mn 0.006, Fe 0.028, Cu 0.006 and Zn 0.109. The concentrations of these metallic elements were calculated as 0.16, 1.5, 0.04, 0.05, and 0.62 ppm respectively in the solution.

Although Brand (1939, 1973) and Brand and Saurwein (1942) have reported the presence of calcium in *Macracanthorhynchus hirudinaceus*, yet the present observations on *P. kashmirensis* have shown that calcium is absent in this species. Absence of calcium is a very unique phenomenon which is difficult to explain. Calcium is reported to be essential for several important processes in the biological system and hence the authors feel that some other element must be acting as a substitute for this element. Regarding other elements it is obvious that concentration of Mn is the highest followed by Zn and Mg.

ACKNOWLEDGEMENTS

The authors are thankful to Mr. Noor-u-Din, Director and Ali Mohd. of USIC, Kashmir University for their assistance in atomic absorption spectrophotometric analysis.

REFERENCES

Brand, T. Von. 1939. Chemical and Morphological observations upon the composition of *Macracanthorhynchus hirudinaceus* (Acanthocephala). *J. Parasitology*, **25** : 329-342.

Brand, T. Von. 1973. *Biochemistry of parasites*. Academic Press, New York.

Brand, T. Von and Saurwein. J. 1942. Further studies upon the chemistry of *Macracanthorhynchus hirudinaceus*. *J. Parasitology*, **28** : 315-318.

Rec. Adv. Fish Ecol. Limn. Eco-conserv.
IV : 116-119

13

The High Altitude Lakes of Kashmir and Their Biological Production

S.M. DAS

When the Himalayas rose some 800,000 years back, the inland sea (Tethys Sea) stretching up to Kashmir was drained off into the Arabian Sea and the Bay of Bengal. This Tethys Sea stretched at one time from the Bay of Bengal through what is now Bengal, Bihar, Uttar Pradesh and Punjab to Kashmir. The rise of the Himalayas not only drained off the sea but left behind fossil remains of marine animals such as Fishes, Brachiopods, Molluscs, etc., which have been collected by the present author from high and dry hill sides in Kashmir Valley at Pahalgam, Zabarwan and Ferozepur Nallah.

The entire valley was raised up to the present height of 1700 m., while the mountain ranges were lifted up to about 5500 m. almost all around the valley, giving it the shape of a bowl. After the sea was drained off, ice-melt, snow-melt and rain filled the bowl with fresh water instead of the old sea water, forming a large cup-shaped lake.

Next came the three glacial periods, when all of north Asia and Europe were iced up from the Arctic southwards; until in the last inter-glacial period fresh flood waters released by melting of ice came into Kashmir valley. The rising of the Pamir mountain-knot in Central Asia caused a high watershed centre from which five finger-like water-shed ranges projected. These today are : Tienshan, Kunlun, Korakoram, the Himalayas and the Hindukush ranges (Das 1966).

The fishes from Central Asia and the Palaearctic region along with other aquatic life came down to Kashmir (Das 1966). They could not move down to the rest of India due to the high barrier of the Pir Panja, Zojila and the western Himalayan ranges. This incursion of Palaearctic fauna took place about 80,000 years back and continued up to 20,000 years ago. This Palaearctic fauna was completely isolated and slowly evolved into the typical Kashmir species. There are today 23 typical endemic Kashmir fishes such as *"Chhruw"*, *"Chuss"*, *"Rupput"*, *"Ramagurun"*, *Aragurun"*, etc., which are found nowhere else in India or the rest of the world (Das 1966, 1970).

Subsequent upheavels due to earthquakes and upthrust on earths crust, caused by northward movement of peninsular India striking against the North

Asian continent, caused islands to be formed in the large Kashmir lake-bowl. Later on, due to lessening of rain, snow and floods, parts of the lake dried up and left behind the scattered high, and low-altitude lakes which are seen all over Kashmir today. Many of the valley lakes, such as Kounsarnag are still interconnected, especially during the floods. Some crater lakes such as Kounsarnag (4500 m.) were formed by water filling up volcanic craters.

According to the Pamir-Kashmir Theory of Das (1966-1970) the origin of Kashmir fishes occurred simultaneously with the last interglacial period due to flow of melting ice-waters from the Pamir shed. Therefore, the endemic fishes of Kashmir belong to the same families or even genera as are found in China, S. Russia and Europe.

THE LAKES

The cold-water lakes of Kashmir are unique in their altitudinal zonation, fauna and flora, their beauty and their biological production. They may be arranged according to the descending order of altitude as follows :

Kounsarnag	(4660 m.)
Shyok	(4500 m.)
Sheshnag	(3800 m.)
Vishensar	(3600 m.)
Gangabal	(3600 m.)
Alapather	(3660 m.)
Kishensar	(3500 m.)
Tarsar	(3100 m.)
Marsar	(3000 m.)
Tulyan	(3000 m.)
Neelnag	(2615 m.)
Manasbal	(1888 m.)
Dal and Nagin	(1860 m.)
Anchar	(1830 m.)
Woolar	(1587 m.)

Many of these lakes were studied for their ecology and organic production in the International Biological Programme (IBP) under the supervision of the present author in the University of Kashmir (1964-1973). The high-altitude lakes above 3000 m. remain ice-covered for almost six months of the year (Kounsarnag-8 months). Productivity in these oligotrophic lakes is minimal and only a few have fishes. Tarsar, Tulyan, Marsar and Neelnag lakes have been recently stocked with trout seed and may yield some good sport fish for the visitor. They have characteristic zoo-, and phyto-plankton (food of fishes) which have been worked out (Das *et al* 1969, 1970).

THE FISHES

The lakes below 2500 m. (Dal, Manasbal, Woolar) are highly productive in plankton, pedon, macrophytes and fish. Even these Valley lakes have their typical fauna and flora, and are the main reservoirs of fish production in Kashmir. About 75 per cent of the fish catch in these lakes is of the coarse exotic mirror carp (*Panjabi Gad*) which has ousted the endemic Kashmir fishes such as *Schizothorax* (*Chhuru*) *Oreinus* (*Chush*) and *Labeo (Roput)*, *Botia (Ramagurun)* and *Nemacheilus (Aragurun)*, which are the fine fishes of Kashmir. Due to high fecundity and adaptibility to eutrophic low oxygenated waters and detrital feeding, the mirror carp has invaded almost all these low-altitude lakes and even the rivers and streams of Kashmir.

The famous trout (*Salmo trutta fario*) of Kashmir is surprisingly absent from all the Kashmir lakes, being found only in clear water fast-running streams such as Ahrabal, Lidder, Sindh, and parts of Jhelum. They do inhabit the lakes also in other parts of the world such as USA, Canada, Australia and New Zealand. This aspect of trout culture has to be looked into in the future for higher production of the trout, the juveniles of which are reared in many hatcheries and then transplanted into streams, and into future lakes as well.

LAKE PRODUCTION

As all human babies all over the world live on milk alone, baby fishes (larvae, fry and fingerlings) feed only on plankton (minute microscopic plants and animals that float about in the water). The phytoplankton (producers) require an optimum amount of sunlight and nutrient salts (nitrates, phosphates and silicates in the main). If the nutrients are low, the plant-plankton is low as in the high-altitude lakes above 3000 m. On the other hand, some low-altitude lakes below 2500 m. have excessive weeds or marcrophytes such as *Myriophyllum, Ceratphyllum, Hydrilla, Salvinia, Potamogeton* and *Chara*. These weeds not only use up the plant nutrients in the water but also shade up sunlight, causing less production of phytoplankton. This in turn causes less zooplankton production as the zooplankters are consumers of phytoplankton. Consequently this leads to starvation or low diet of the fish larvae, fry and fingerlings, 50 to 80 per cent of these endemic fish juveniles dying out in the lakes. Then again, pollution kills off large numbers of these juveniles, especially in Dal lake.

The endemic fine fish juveniles feed on plankton only, and the adults feed on aquatic weeds, some aquatic animals and algae; while the mirror carp (even juveniles) can feed on putrefying organic matter (which is increasing in these low-altitude lakes) and also bottom detritus. This has caused overcrowding of the mirror carp and the decline of the endemic fine fishes of Kashmir lakes.

The remedies for this decline may be summarised as under :

(1) Removal of excess weeds from the lakes annually by grass cutters (now in action) or by chemicals which would kill the weeds but not the fish.

(2) Weeds can be reduced by introduction of the Chinese grass carp (*Ctenopharyngodon idella*) which eats only the aquatic weeds.

(3) Prevent pollution of Dal lake by special arrangements to pipe off the sewage and excreta, which falls into the lake through the flush systems of over 2000 house-boats all over the lake (Nehru Park, Nishat basin, Bod Dal, and Nagin basin). This sewage and sullage may be collected and converted easily into fertilizers.

(4) Lake encroachment by land-cultivation arms into the lake and inshore hanging vegetable gardens should be prohibited.
(5) Mirror carp should be intensively fished, while size regulation for the endemic fish catch and also closed seasons should be enforced.
(6) Induced breeding and seeding of lakes by resultant fry and fingerlings must be done.

Unless these measures are adopted, the famous endemic Kashmir fishes (viz. *Schizothorax, Oreinus, Labeo dyocheilus*) will become extinct in the next half century or so; even as the mirror carp (*Cyprinus carpio*) has ousted 80 per cent of the endemic fishes in the past 25 years, after it was introduced in the Kashmir lakes; and the famous Mahaseers (*Tor tor* and *T. putitora*) are declining in Kumaun lakes.

Besides fish production, the low-altitude Kashmir lakes are famed for the lotus stem (*nadru*) for human consumption and some zones for cultivation of *Trapa* (*Singhara*). The lakes of Kashmir form the best riches for the Kashmir people, supplying them with water, fish, edible plants and floating gardens.

The maximum benefit of the lakes to the Kashmiris is the high tourist trade from India and abroad, as the lakes (and streams) afford a large scope for boating, yachting, water skiing, angling for trout and endemic fishes, as well as the marvellous sight-seeing and landscape-painting offered to the more sensitive eyes. The recent air-bus service from Delhi to Srinagar will surely provide opportunities to more people to enjoy the lakes of Kashmir, both for study and recreation.

REFERENCES

Das, S.M. 1966. A revision of fishes from Kashmir Province. *Kashmir Sci.*, **2** (1-2) : 64-76.

Das, S.M. 1966. Palaearctic elements in the fauna of Kashmir. *Nature*, **212** : 1-6.

Das, S.M. 1967. Recent advances in fish ecology. Presidential Address Biological Sciences Section, *Nat. Acad.* Sci., 1-10.

Das, S.M. 1970. Ecology of Dal lake, Kashmir. *Polish I B P Ecology Symposium Report*, 44-60.

Das, S.M. 1970. The Pamir-Kashmir Theory of the origin and evolution of the ichthyofuna of Kashmir Valley. *Ichthyologica*, **10** : 21-27.

Das, S.M. and Shabir Hussain 1967-69. Studies on induced breeding by mammalian hormone injection in Kashmir fishes : (i) *Salmo gairdneri gairdneri* and *Salmo trutta fario*; (ii) *Cyprinus carpio specularis*; (iii) *Schizothorax esocinus. Kashmir Sci.*, **4** : 206-215; **5** : 78-84; **6** : 85-90.

Das, S.M. and Subla, B.A. 1963. The ichthyofauna of Kashmir, Part I : History, topography, origin, ecology and general distribution. *Ichthyologica*, **2** (1-2) : 87-106.

Nath, Surendra 1992. Ichthyofauna of Jammu & Kashmir State (India) with remarks on the ichtyhygeography of the region. *Rec. Adv. Fish Ecol. Limn. Eco-conserv.*, **11** : 1-32.

Zutshi, D.P. *et al.* 1971. Limnology of high altitude Kashmir lakes. *Intern. Limnol. Congress, Leningrad.*

Zutshi, D.P. *et al.* 1978. Limnological studies on Dal lake, Kashmir. *J. Inland Fish. Soc. India*, **11** (1) : 12-21.

REFERENCES

(4) Lake encroachment by land-cultivation arms into the lake and inshore hanging vegetable gardens should be prohibited.
(5) Mirror carp should be intensively fished, while size regulation for the endemic fish catch and also closed seasons should be enforced.
(6) Induced breeding and seeding of lakes by resultant fry and fingerlings must be done.

Unless these measures are adopted, the famous endemic Kashmir fishes (viz. *Schizothorax, Oreinus, Labeo dyocheilus*) will become extinct in the next half century or so; even as the mirror carp (*Cyprinus carpio*) has ousted 80 per cent of the endemic fishes in the past 25 years, after it was introduced in the Kashmir lakes; and the famous Mahaseers (*Tor tor* and *T. putitora*) are declining in Kumaun lakes.

Besides fish production, the low-altitude Kashmir lakes are famed for the lotus stem (*nadru*) for human consumption and some zones for cultivation of *Trapa* (*Singhara*). The lakes of Kashmir form the best riches for the Kashmir people, supplying them with water, fish, edible plants and floating gardens.

The maximum benefit of the lakes to the Kashmiris is the high tourist trade from India and abroad, as the lakes (and streams) afford a large scope for boating, yachting, water skiing, angling for trout and endemic fishes, as well as the marvellous sight-seeing and landscape-painting offered to the more sensitive eyes. The recent air-bus service from Delhi to Srinagar will surely provide opportunities to more people to enjoy the lakes of Kashmir, both for study and recreation.

REFERENCES

Das, S.M. 1966. A revision of fishes from Kashmir Province. *Kashmir Sci.*, **2** (1-2) : 64-76.

Das, S.M. 1966. Palaearctic elements in the fauna of Kashmir. *Nature*, **212** : 1-6.

Das, S.M. 1967. Recent advances in fish ecology. Presidential Address Biological Sciences Section, *Nat. Acad.* Sci., 1-10.

Das, S.M. 1970. Ecology of Dal lake, Kashmir. *Polish I B P Ecology Symposium Report*, 44-60.

Das, S.M. 1970. The Pamir-Kashmir Theory of the origin and evolution of the ichthyofuna of Kashmir Valley. *Ichthyologica*, **10** : 21-27.

Das, S.M. and Shabir Hussain 1967-69. Studies on induced breeding by mammalian hormone injection in Kashmir fishes : (i) *Salmo gairdneri gairdneri* and *Salmo trutta fario*; (ii) *Cyprinus carpio specularis*; (iii) *Schizothorax esocinus. Kashmir Sci.*, **4** : 206-215; **5** : 78-84; **6** : 85-90.

Das, S.M. and Subla, B.A. 1963. The ichthyofauna of Kashmir, Part I : History, topography, origin, ecology and general distribution. *Ichthyologica*, **2** (1-2) : 87-106.

Nath, Surendra 1992. Ichthyofauna of Jammu & Kashmir State (India) with remarks on the ichtyhygeography of the region. *Rec. Adv. Fish Ecol. Limn. Eco-conserv.*, **11** : 1-32.

Zutshi, D.P. *et al.* 1971. Limnology of high altitude Kashmir lakes. *Intern. Limnol. Congress, Leningrad.*

Zutshi, D.P. *et al.* 1978. Limnological studies on Dal lake, Kashmir. *J. Inland Fish. Soc. India*, **11** (1) : 12-21.

Index